Hidden In Plain Sight 8

Andrew Thomas studied physics in the James Clerk Maxwell Building in Edinburgh University, and received his doctorate from Swansea University in 1992.

His *Hidden In Plain Sight* series of books are science bestsellers.

Also by Andrew Thomas:

Hidden In Plain Sight
The simple link between relativity and quantum mechanics

Hidden In Plain Sight 2
The equation of the universe

Hidden In Plain Sight 3
The secret of time

Hidden In Plain Sight 4
The uncertain universe

Hidden In Plain Sight 5
Atom

Hidden In Plain Sight 6
Why three dimensions?

Hidden In Plain Sight 7
The fine-tuned universe

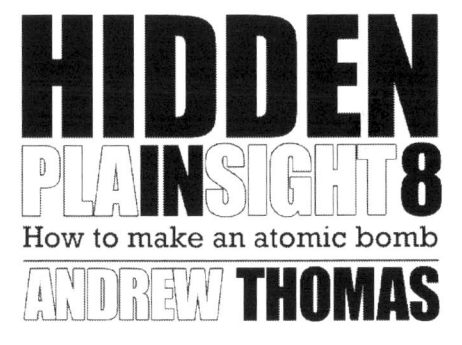

HIDDEN PLAINSIGHT 8
How to make an atomic bomb
ANDREW THOMAS

**AGGRIEVED
CHIPMUNK
PUBLICATIONS**

AGGRIEVEDCHIPMUNK.WORDPRESS.COM

Hidden In Plain Sight 8

Copyright © 2017 Andrew D.H. Thomas

All rights reserved.

ISBN-13: 978-1548577810
ISBN-10: 1548577812

CONTENTS

Prologue

1 Our friend the atom   1
    The Gilbert Atomic Energy Lab
    Operation Plowshare
    Project Orion

2 Faint fairy lights   17
    Alpha and beta
    The alchemists
    Nuclear reactions
    The power of the atom
    Beta radiation
    The weak interaction

3 Splitting the atom   41
    Artificial radioactivity
    The discovery of fission
    The nuclear mousetrap

4 The chain reaction   57
    The New World
    The letter from Einstein
    Oppenheimer
    Los Alamos
    The *Los Alamos Primer*

5   The critical mass   75
   The ratio of volume to surface area
   Spherical coordinates
   The build-up of neutrons
   The derivative
   To cut a long story short …

6   Oak Ridge   87
   The secret city
   The largest building in the world
   The Calutron Girls
   Mail-order uranium

7   Plutonium   107
   The nuclear reactor
   The world's first nuclear reactor
   The Demon Core

8   Detonation   123
   The sex of the bomb
   Implosion
   Trinity

Appendix: Calculating the critical mass
   Creating the equation
   The genius of Mr. Wolfram
   Solving the equation
   Calculating the critical mass

PROLOGUE:
THE Nth COUNTRY EXPERIMENT

The year is 1964.

Dave Dobson was a physics student, excited to have just received his PhD. Dobson was a bright guy, and a science enthusiast. He had a good general knowledge about several fields of physics, but his knowledge was nothing spectacular.

To all intents and purposes, Dobson was Mr. Average.

This made it all the more surprising when Dobson received a telephone call from the renowned nuclear physicist Edward Teller. Teller invited Dobson to come to Washington D.C. for an interview for a job in one of the country's leading nuclear research facilities. Dobson was amazed and, frankly, wondered if they had got his name mixed up with someone else. He later described everything he knew about nuclear chain reactions at that time: "I had seen an exhibit with a model of a chain reaction made up of mousetraps and ping pong balls."

Dobson met Teller and they spent an evening together. Teller quizzed Dobson to discover everything he knew about nuclear weapons, and Dobson honestly replied that he knew nothing more than any other amateur science enthusiast. To be frank, Dobson admitted he knew absolutely nothing at all about nuclear weapons.

"Great", replied Teller. "You will be perfect for the job".

At that time – the early 1960s – the Pentagon was extremely concerned about the possible proliferation of nuclear weapons to other countries. Only the United States, the Soviet Union, Britain, and France possessed nuclear weapons. Britain had been the third country, France had been the fourth. Which country would be the fifth country to possess nuclear weapons? Or the sixth country? Where would it all end? Which country would be the "Nth country"?

In order to shed some light on that question, the Pentagon had started a top secret project called the "Nth Country Experiment". Dave Dobson would be one of two physics students who had been selected to work on the project.

The worry was that so much information about how to make an atomic bomb had been published in the popular press. While no classified secrets had been leaked publically, there was plenty of more general information freely available to an amateur enthusiast. Given the availability of that general information, would it be possible for a rogue state, employing nothing more than a few physicists of very average ability, to create a nuclear weapon?

Basically, the goal of the Nth Country Experiment was to determine how difficult (or easy) it was for a bunch of amateurs to make an atomic bomb.

This, then, was to be Dave Dobson's first job straight out of university. Dobson was given a desk in a corner of a laboratory in the Livermore Radiation Laboratory in California. He was introduced to another amateur physics enthusiast, Bob Selden, a 28-year-old soldier who would be working with him. Neither man had any nuclear expertise whatsoever.

Their job was to imagine they were working for a rogue state (named the "Nth Country"). They were handed a document explaining their task. Here is the first paragraph of the document:

PROLOGUE

> *The purpose of the so-called "Nth Country Experiment" is to find out if a credible nuclear explosive can be designed, with a modest effort, by a few well-trained people without contact with classified information. The goal of the participants should be to design an explosive with a militarily significant yield. A working context for the experiment might be that the participants have been asked to design a nuclear explosive which, if built in small numbers, would give a small nation a significant effect on their foreign relations.*

The first thing which needed to be done was to obtain security clearance for Dobson (Selden had already been cleared because of his military background). Clearance was necessary because it was against the law to design nuclear weapons without security clearance.

Dobson and Selden would have access to none of the secret research material at the laboratory, but they would have access to the general library and its publically-available material. They were told that the imaginary "Nth Country" they were working for would have "more resources than Ghana, but less than an industrialized nation." They should assume they had access to good machinists able to shape uranium and plutonium, and also access to some engineers experienced with conventional explosives. Apart from that, they were given no more instructions on how they should proceed. They were on their own.

In the library, Selden found a book about the Manhattan Project, the U.S. project which developed the atomic bomb. According to Dobson: "It gave us a road map. But we knew there would be important ideas they'd deliberately left out because they were secret. This was one of the things that produced a little bit of paranoia in us. Were we being led down the garden path?"

If Dobson and Selden wanted to perform an experiment, perhaps with high explosives, they had to describe their desired experiment in great detail. A team of experts then calculated the result of the experiment and passed the result back to Dobson and Selden.

Ironically, one of the most useful sources of information came from President Eisenhower's "Atoms for Peace" programme, the motivation for which was to encourage the use of non-military nuclear power around the world. Atoms for Peace was just one example of the enthusiasm at the time for nuclear energy, the result of which was to distribute a vast amount of technical information into the public domain.

After two-and-a-half years, in 1966, Dobson and Selden had finally finished their design. According to Selden: "We produced a short document that described precisely, in engineering terms, what we proposed to build and what materials were involved. The whole works, in great detail, so that this thing could have been made by Joe's Machine Shop downtown."

However, after they had presented their document, everything went very quiet, and Dobson and Selden were not informed of the outcome. They presumed that they had failed in their task. So, one day in Livermore when they met a senior researcher, Jim Frank, they asked him why things had gone so quiet. Did he have any knowledge of the outcome of their experiment? Yes, said Frank, he knew what had happened. The reason why everyone had gone quiet was because they had realised that if a bomb had been constructed precisely according to their detailed instructions, it would have exploded with the same order of magnitude as Hiroshima.

1

OUR FRIEND THE ATOM

In 1870, the French science fiction writer Jules Verne wrote the classic adventure novel *Twenty Thousand Leagues Under the Sea*. It is the story of Captain Nemo who terrorised the world's shipping using his mysterious submarine called the *Nautilus*.

As usual, Jules Verne showed his amazing talent as a futurist, predicting technological advances many years before that technology came to exist. In this case, the submarines of the 19th century were very primitive vessels, so Jules Verne's description of a craft which could stay submerged for several days proved to be prophetic. That was because, in 1954, the real-life Nautilus was launched. And it could, indeed, travel underwater for weeks or months at a time.

The *USS Nautilus* – named after the craft in Jules Verne's book – was the world's first nuclear-powered submarine. In a nuclear-powered submarine, heat generated by a small nuclear reactor turns water into steam which, in turn, powers the propeller of the submarine. Unlike a diesel engine, a nuclear reactor requires no air to operate. As a result, nuclear-powered submarines are capable of staying at sea for many months without needing to refuel, and can travel huge distances under water.[1] To prove the point, in 1958 the real-life Nautilus took four days to complete the first submerged voyage under the North Pole.

The following photograph shows the USS Nautilus arriving in New York City in 1958:

America in the 1950s was an optimistic time, and nuclear power was viewed very differently from how it is viewed today. The almost unlimited energy produced by the atom gave the promise of cheap, compact power sources which could be used for almost any application. The success of the USS Nautilus inspired ideas for other nuclear-powered forms of transport. One of the most outrageous suggestions included the world's first nuclear-powered car, the Ford

---

[1] Note that the "Twenty thousand leagues" of Jules Verne's book refers to a huge distance travelled – not depth under water, as twenty thousand leagues is twice the circumference of the Earth.

Nucleon which was proposed by the Ford Motor Company in 1958. In the rear of the car was to be a small nuclear reactor which would allow the car to travel 5,000 miles without refuelling (unfortunately the car never got off the drawing board: such a small nuclear reactor did not exist, and the required shielding to protect the passengers would have made the car too heavy).

It was widely-believed that the electricity produced by nuclear power would be so cheap that it would make more sense to provide it for a flat fee – or even give it away for free – rather than installing expensive meters. This optimism was summed-up in a 1954 speech by Lewis Strauss who was the chairman of the United States Atomic Energy Commission:

> *Our children will enjoy in their homes electrical energy too cheap to meter. It is not too much to expect that our children will know of great periodic regional famines in the world only as matters of history, will travel effortlessly over the seas and under them and through the air with a minimum of danger and at great speeds, and will experience a lifespan far longer than ours, as disease yields and man comes to understand what causes him to age.*

It is believed that Strauss was referring to power produced by nuclear fusion, which would provide almost unlimited power from readily-available hydrogen. Indeed, if nuclear fusion research proves successful then Strauss's prediction of too-cheap-to-meter energy might yet come true.

In 1954, feeding off the excitement caused by the launch of the Nautilus, Walt Disney released the movie *Twenty Thousand Leagues Under the Sea* starring Kirk Douglas and James Mason as Captain Nemo. To support the release of the movie, an episode of the Disney television series

Disneyland was devoted to examining the wonders of nuclear power. Continuing the optimistic tone of the age, the episode was called *Our Friend The Atom*.

The *Our Friend The Atom* episode is an excellent introduction to nuclear physics and it is available on YouTube:

http://tinyurl.com/movieatom

The programme starts with Walt Disney standing beside a scale model of the USS Nautilus (including a clip from the Disney movie *Twenty Thousand Leagues Under the Sea* featuring the fictional Nautilus). We are then shown around Disney's "Atomic Energy Lab" which features some very nice wood-panelling and some "scientists" in white coats looking down optical microscopes (presumably at "atoms"?). It is rather bizarre to hear Walt Disney say: "As you can see, we found ourselves deep in the field of nuclear physics."

We are then introduced to Heinz Haber, a physicist who worked for the Luftwaffe in World War Two. After the war, he was taken to America together with Wernher von Braun to keep his expertise out of the hands of the Soviet Union. Haber went on to provide valuable contributions to the NASA space programme.

Haber then introduces some Disney animated sequences which are used to illustrate various issues associated with nuclear energy. For example, in a tale from *The Arabian Nights*, nuclear energy is compared to a genie in a lamp which has the power to do good or evil, but can never be trapped back inside the bottle from which it came.

Heinz Haber then proceeds to tell the story of the discovery of the atom, starting in ancient Greece with Democritus who suggested that matter was formed of indivisible small hard balls. Democritus could never provide any evidence to support his atomic theory. However, the story then moves to consider the work of the 19th century

English chemist John Dalton who proposed the atomic theory of chemistry. Dalton realised that the combination of chemical elements (such as hydrogen and oxygen) to form chemical compounds could be explained if the elements were formed of atoms. This was the first truly scientific theory of the atom. According to Dalton, when atoms of different elements joined together, compounds were formed, e.g., two hydrogen atoms can combine with one oxygen atom to form water.

Haber then considers advances in atomic theory through the 19th and early 20th centuries. Slowly, Haber builds-up the model of the atom.

Atoms are composed of a tightly-packed nucleus surrounded by orbiting electrons. The nucleus is made out of positively-charged *protons* and electrically-neutral *neutrons*. Because protons have positive electric charge, their tendency is to repel each other. The neutrons, therefore, have to act like glue to hold the protons together and to stop the nucleus from flying apart. The negatively-charged electrons are held in their orbits by the electrical attraction to the positively-charged protons (opposite charges attract).

The following diagram shows a carbon atom. It is composed of six electrons orbiting a nucleus which is composed of six protons and six neutrons:

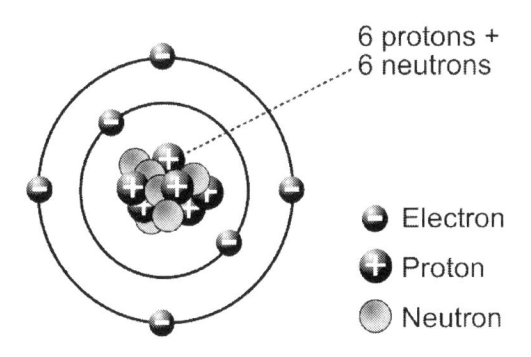

In this book we are only going to be interested in the nucleus of the atom. The study of the nucleus of the atom is called *nuclear physics*. In his book *The Making of the Atomic Bomb*, Richard Rhodes reveals the key reason why it is the nucleus of the atom which provides the energy for nuclear explosives: "Nuclear physics, the study of the nucleus of the atom, is where most of its mass – and therefore its energy – is concentrated."

A Disney book was published in association with the television programme, also called *Our Friend The Atom*. The book was written by Heinz Haber who hosted the television programme. Here is a photo of me with a copy of the book in good condition which I bought off eBay:

When my copy of the book arrived I was delighted to find the pages were pristine white, completely unfaded, and I got the impression the book had not been opened for over fifty years. This impression was strengthened when I opened the book and found a cutting from the *Daily Telegraph* newspaper dated April 13th 1961 announcing that the Soviet astronaut Yuri Gagarin had returned safely from the first manned space flight, having completed an orbit of the Earth.

I was even more delighted to find hand-written notes (on perfect white paper) detailing contemporary hydrogen bomb tests and other atomic research. The following photograph shows the notes I found in my book:

The book had clearly belonged to a science enthusiast. It is an example of the interest and excitement of the general public about atomic energy in the 1950s and early 1960s. I was delighted to find these hidden secrets – I guess sometimes you just get lucky.

Our Friend The Atom was obviously aimed at a young audience, but it was a superb project. It is a shame to think that such a high-level educational programme about nuclear energy would never be aimed at young children today. It is also a shame that we have lost our optimism and enthusiasm about the potential contained within the atom. In this book, I hope to restore some of that optimism.

The Gilbert Atomic Energy Lab

Our Friend The Atom was not the only example of 1950s nuclear education which was aimed at children. Take a look at the following toy. The boy on the lid of the box looks thrilled. No wonder! He has just been given the Gilbert Atomic Energy Lab for his birthday and he is busy doing experiments with radioactive material:

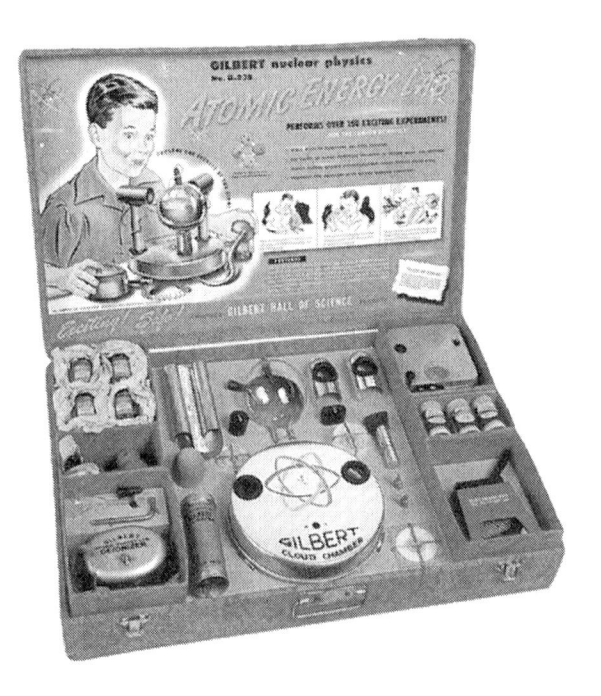

According to its Wikipedia page, the Gilbert Atomic Energy Lab has been called "The world's most dangerous toy", which is extremely unfair and – frankly – completely inaccurate. I would prefer to call it "The world's coolest toy ever".

Alfred Gilbert, the toy maker who released the Atomic Energy Lab, believed that his toys held the key to building a "strong American character", and his toys always included some educational quality. But the Atomic Energy Lab contained genuine radioactive ore samples. So was the Atomic Energy Lab a safe toy?

Well, in assessing whether it was safe, it is important to realise that we are all constantly being irradiated by a low-level background radiation. This radiation may come from the rocks and soil which surround us, or even from the plants which we eat – particularly bananas, which are surprisingly radioactive. This has resulted in the creation of a measure of radiation called the *banana equivalent dose* (or BED). One BED is approximately equal to 1% of a day's worth of background radiation (which would therefore be equal to 100 BEDs). The level of this background radiation is low, but over a year it all adds up. It is true that the radioactive ore samples in the Atomic Energy Lab were considerably more active than the constant background radiation, but you would not be exposed to those samples for long. In evaluating the safety of the Atomic Energy Lab it is important to realise that "safe" is a relative concept.

It is perhaps a shame that nowadays the concept of giving children an atomic energy lab seems bizarre. It exemplifies the public misconceptions about the relative risks of radioactivity. We have lost sight of the positive contributions of radioactivity to our world, from power generation to cancer radiotherapy. I am sure Alfred Gilbert's laboratory was a wonderfully educational toy.

Later in this book we will be considering my own experiments with radioactive substances when I have fun by creating my own version of the "Gilbert Atomic Energy Lab" using a Geiger counter and some highly-radioactive uranium ore. I can assure you that I do live to tell the tale!

Operation Plowshare

In the 1950s, there was also a much more positive attitude towards nuclear explosives. Yes, it was the case that the atomic bomb had been recently used in warfare, but the general belief was that its use had prevented the loss of the lives of hundreds of thousands of American servicemen during an invasion of the Japanese mainland. It was also believed that the awesome destructiveness of the bomb would lead to the end of mass warfare as countries realised the futility of "Mutually Assured Destruction", or MAD.

As part of that more positive approach to nuclear explosives, the U.S. government started Operation Plowshare to investigate the use of nuclear explosives for peaceful construction purposes. The name "Plowshare" came from the Biblical quote "They shall beat their swords into plowshares", indicating a desire to convert weapons into tools. Proposed uses of nuclear explosives included the excavation of large amounts of earth and rock, creating artificial harbours, widening the Panama canal, cutting paths through mountains for highways, and creating underground caverns for mining and storage.

The largest nuclear test of Operation Plowshare was the Sedan test which took place on 6th July 1962. The aim of the test was to examine the feasibility of using nuclear explosives to excavate large amounts of earth and rock. The nuclear explosive was buried 194 metres underground, and the explosion displaced twelve million tons of rock. The resultant crater is called the Sedan Crater and it is the largest man-made crater in the United States. Over 10,000 visitors a year now visit the site on free monthly tours.

The Sedan test took place at the Nevada Test Site, which is a 1375-square-mile area of empty desert which is located only sixty miles away from Las Vegas. The first atomic test at the site took place in 1951, and over the next four decades it hosted a further 928 tests giving the area the nickname of "the most bombed place on Earth". Even today, if you search for "Nevada National Security Site" on Google Maps you will find the pock-marked area of desert, with each crater representing a nuclear explosion. It is even possible to see the enormous Sedan Crater at the north of the site.

In 1951, Las Vegas was a struggling small town with a population of fewer than 25,000 which was looking to boost its profile. When atomic bomb testing started at the nearby test site, the Las Vegas Chamber of Commerce promoted the blasts as a tourist attraction, handing out leaflets giving the dates of the detonations. Hotels built north-facing penthouses with the best views of the atomic mushroom clouds.

The following photograph was taken in Las Vegas and shows the mushroom cloud of a nuclear explosion at the Nevada Test Site rising high above the casinos in the foreground:

Project Orion

One of the most remarkable proposed uses for nuclear explosives in this period was Project Orion. The aim of Project Orion was to design a spacecraft which was to be propelled by a sequence of thousands of nuclear bombs exploded in succession at the rear of the craft. The craft was to be protected by a 1,000-ton steel plate (called a *pusher plate*) and shock absorbers to protect the crew from the crushing acceleration (though that would not be a problem if the craft was unmanned). Radioactive nuclear fallout would not be a problem if the propulsion system was restricted to being only used in space. So, although the Project Orion proposal might sound absurd, experiments at the time showed that it would actually be a viable means of space travel.

Project Orion was described in detail in the BBC documentary *To Mars by A-Bomb: The Secret History of Project Orion*. The video is absolutely fascinating and is well-worth watching. The video is available at the following link:

http://tinyurl.com/projectorionmovie

If all else fails, the following link to the video should always work:

http://tinyurl.com/projectorionbackup

In the video, the physicist and author Arthur C. Clarke emphasizes that the idea behind Project Orion was not crazy: "After all, every time you get into a motor car you are being driven around by means of a series of rapid explosions."

The following image shows a NASA artist's impression of the Orion spacecraft with Saturn drawn in the background. The image is taken about four milliseconds after the explosion of a nuclear propellant charge, showing the blast hitting the pusher plate at the rear of the craft. The line drawing underneath is another NASA diagram showing the key components of the Orion craft, including the pusher plate and the multiple shock absorbers:

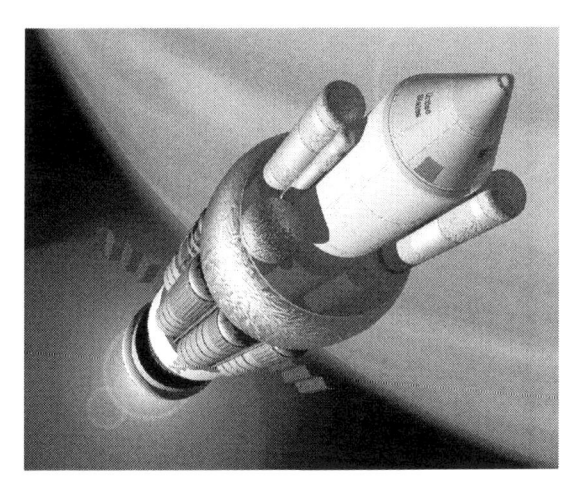

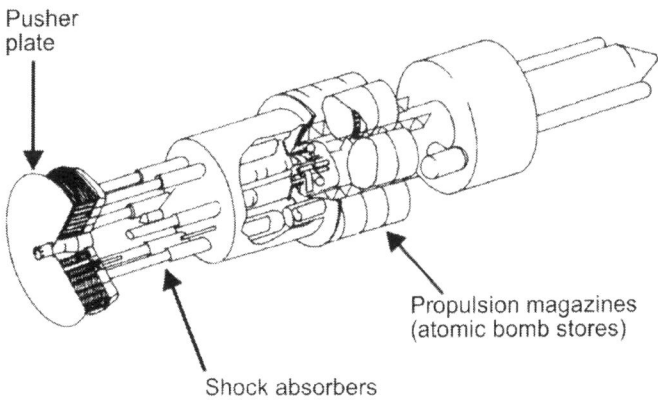

Pusher plate

Propulsion magazines (atomic bomb stores)

Shock absorbers

As explained in the BBC video, if the spaceship was of standard size (e.g., just a few tons in weight) then the force of the exploding bombs would pulverise the craft. The solution was to make the spacecraft big: a few thousand tons in weight. According to Newton's second law of motion, the resultant acceleration would be greatly reduced due to the immense mass. Not only would this save the structure of the craft, it would also make the acceleration endurable for the on-board travellers. The plan was therefore to make an enormous craft, the size and weight of an ocean liner. To this day, the Orion technology remains our only method for moving extremely large payloads around the Solar System because of the huge energy advantage of nuclear explosives over chemical propellants. It has been said that: "To this day, Orion is still the only feasible means of interstellar travel, both robotic and manned, that could actually be built with current technology and knowledge."[2]

As a result of the necessity for large mass, the spacecraft would be built using a radically new philosophy. The spacecraft would be constructed out of steel, resembling a building more than a spacecraft. In fact, it would be more like a hotel – or an ark. It would have the potential to carry a large population out of the Solar System to colonise nearby planetary systems. As Freeman Dyson – one of the physicists who worked on Project Orion – says in the BBC video: "Establishing human colonies was certainly part of our plan".

But surely there could never be a pressing need for such an ark. Why would humanity ever need to leave the Earth? Why would humanity need to colonise distant planets? Well, there is potentially one very good reason: a large asteroid or comet might appear on a collision course with the Earth

---

[2] Paul Gilster, *Project Orion: A Nuclear Bomb and Rocket – All in One*, http://tinyurl.com/centauridreams

threatening the mass-extinction of all life on the planet. In that situation, Orion might provide our only chance of survival.

If an asteroid on a collision course is detected a long way from Earth, there would be time to use "kinetic impact" – ramming a spacecraft into the asteroid – in order to produce a slight deflection to the course of the asteroid. Because of the great distance left on its course, that small deflection might be enough to cause the asteroid to miss the Earth. Time is of the essence when it comes to asteroid deflection.

Unfortunately, we would probably not have a long-enough warning period to launch such a spacecraft. In 2014, a comet was detected on a collision course with Mars – just 22 months before it crashed into the planet. According to Dr. Joseph Nuth, a NASA researcher, that would not have been nearly enough time to launch a rocket had the comet been on a collision course with the Earth. It takes five years to design and launch a spacecraft.

According to the BBC documentary, Orion might be the best technology to intercept and deflect an Earth-bound asteroid. Johndale Solem, a physicist at Los Alamos National Laboratory, said that "Orion provides such an advantage in speed over chemical propellants that it seems that interception could take place in a much shorter timescale, and consequently the deflection could take place further away so that it would be easier to make such an object miss the Earth."

Once intercepted, nuclear explosives could also be deployed either close to the asteroid to deflect it, or they could be used on the surface or beneath the surface of the asteroid in order to destroy it. The public perception of nuclear explosives might be very negative, but one day we might need to rely on them to save our planet.

I hope this introductory chapter has convinced you that nuclear explosives are not necessarily evil and maybe one day – in a world without war – they may prove to be of tremendous benefit to humanity. After all, dynamite has been a great boon to the construction and demolition industries, and we do not consider there to be anything fundamentally evil about dynamite.

A nuclear explosion is just an astonishing natural phenomenon. How we harness the power of that natural phenomenon represents a challenge – and an opportunity – for humanity.

Let us now start the long story of the development of the atomic bomb …

2

FAINT FAIRY LIGHTS

Marie Curie looked at the test tube in delight. The year was 1902. Marie was thirty one years old, and working in the laboratory of her husband Pierre at the School of Physics and Chemistry in Paris. The test tube contained radium, a new element which Marie and Pierre had just discovered. To her delight, she noticed the test tube was glowing with a beautiful green colour, an effect she innocently called "faint fairy lights". Even when Marie closed her eyes she could still see the magic colours as the rays penetrated her eyelids as though they were not there.

In the evenings, she would often sit with Pierre in the laboratory they shared and gaze in wonder at the eerie blue-green glow of their new discovery which illuminated the entire room. Marie gave a name to these emitted rays: *radioactivity*. In this chapter we will be considering radioactivity in detail.

Six years earlier, the French physicist Henri Becquerel had first discovered the radioactive properties of uranium. Becquerel was investigating the effect of sunlight on photographic plates when he was frustrated by a cloudy day in Paris. Unable to perform his experiment due to the lack of sunlight, Becquerel put away his photographic plate into a drawer – which also happened to contain a nugget of uranium.

The next day, when Becquerel took the plates out of the drawer, he was surprised to find that the uranium had left clear imprints on the plates as though they had been exposed to a bright light. This was the first detection of radioactivity, and Becquerel's cloudy day in Paris stands as a clear example of the role of good luck in scientific discovery.

The discovery of radioactivity is presented in animated form at the 19:53 minute mark of Disney's *Our Friend The Atom*. This link takes you there directly:

http://tinyurl.com/radioactivediscovery

Becquerel did not take his work any further, so it was left to Marie and Pierre Curie to pursue this mystery.

Marie discovered that the rays were purely a product of uranium on its own – not due to any chemical reaction with any other element. Therefore, Marie had the insight that the rays were a fundamental property of uranium atoms. It was also clear that there was considerable energy being produced purely from the atom. This was the first observation of the potential of atomic power. Marie Curie's "faint fairy lights" were going to light up the world.

But if uranium was radioactive, might not other elements be radioactive as well? This was the question which was to lead to the discovery of radium. Marie tested all eighty known elements for radioactivity, and many other substances as well. Eventually, Marie found one particular mineral called pitchblende – a by-product of mining – which appeared to be far more radioactive than uranium. If radioactivity really was generated from the atoms of a single element then there had to be some new and mysterious element in the pitchblende compound. But it would be no easy task to refine the mystery element. The Curies obtained seven tons of pitchblende from the mines of Bohemia, a pile of black rubble which filled their laboratory. The rubble had to be laboriously crushed by hand in a pestle and mortar. But after four years of hard labour, the Curies managed to extract one tenth of a gram of a new element – radium – from the pile of rubble.

The discovery of radium proved to be an international sensation. It appeared to be a new, constantly-glowing miraculous source of energy – ideal for the luminous paint of watch dials. Radium was also viewed as a potential cure for numerous diseases, including throat medicines and cough cures, and you could even buy radium toothpaste, condoms, and suppositories.

Marie and Pierre were interested in the medical applications of their new discovery, and were even willing to sacrifice their own health in their research. At one time, Pierre strapped a test tube filled with radium to his arm for ten hours, creating a lesion on his arm. Bizarrely, Pierre was thrilled – and for good reason. If radiation could damage healthy tissue, then it should also be able to destroy diseased tissue. This gave the Curies the idea to use radioactivity to treat cancer – the first use of radiotherapy. Radiotherapy represents one of the ways in which radiation has provided benefits to humanity.

The declining health of Pierre was noted when Edward Rutherford came to visit the Curies in Paris in the summer of 1903. To celebrate the award of Marie's doctorate, a party was held in their garden. Pierre brought out a glowing tube of radium which impressed Rutherford greatly, describing it as "a splendid finale to an unforgettable day". However, the light was bright enough to reveal Pierre's hands to Rutherford, who noted that they were inflamed and painful due to exposure to radioactivity.

In April 1906, Pierre was crossing the busy intersection of Rue Dauphine near the Seine when he fell under a horse-drawn carriage. He narrowly missed being trampled by the horses as the driver attempted to stop the carriage, but the carriage was carrying six tons of military equipment and kept slowly rolling. The back wheels of the carriage crushed Pierre's head, killing him instantly.

Marie was devastated by the loss of her husband and colleague. The Sorbonne (the University of Paris) awarded Pierre's professorship to Marie, making Marie the university's first female professor in its 650-year history.

By 1911, Marie was feeling more content about her personal situation, and there were three reasons for her improved state of mind. The first reason for Marie's happiness was that she had just received a telegram confirming that she had won her second Nobel Prize. That award of the 1911 Nobel Prize in chemistry, together with her 1903 award in physics for the discovery of radioactivity, made her the first person to win two Nobel prizes.

Secondly, Marie was invited to be a participant at the first International Solvay Conference in Brussels, which was a meeting of the greatest minds in physics. Attendees included Albert Einstein, Max Planck, and Ernest Rutherford. The following photograph shows the attendees, with Marie Curie the second seated person from the right:

FAINT FAIRY LIGHTS

The third reason for Marie's happiness was the presence of her lover, Paul Langevin, who was standing next to Einstein at the extreme right of the Solvay photograph. Unfortunately, Langevin also happened to be a married man. When Langevin's wife saw the photograph of Marie and Paul together at the Solvay Conference she told the press of the affair and it quickly became a national scandal with lurid stories in the newspapers. At one point, an angry mob formed outside Marie's house and threw stones through her window. Einstein wrote a letter of support to Marie regarding her treatment by the press: "If the rabble continues to be occupied with you, simply stop reading that drivel. Leave it to the vipers it was fabricated for."

Paul Langevin, defending his honour, challenged the journalist of the newspaper to a duel. The two duellists met at eleven o' clock in the morning. Each took their pistol, and paced-out twenty-five yards. But when they turned to face each other, the journalist fired into the ground. Explaining his action later, he said: "Paul Langevin has a reputation as a scientist. However grave may be the errors made by Langevin in his domestic life, I did not wish to deprive French science of a precious brain." On seeing his adversary's action, Langevin also fired into the ground, saying: "I am not an assassin".

Langevin returned to his wife, and the outrage over the affair was brief. But the incident gives us an insight into the non-conformist life of Marie Curie, and reminds us of the challenging culture in which she had to work.

Marie Curie died at the age of 66 from exposure to radiation. The bodies of Marie and Pierre Curie were interred in the Panthéon in Paris. To this day, Marie's original documents remain so highly radioactive that they are kept in lead-lined boxes and can only be read whilst wearing protecting clothing.

Alpha and beta

As has just been described, Marie Curie explained that radioactivity had to be a property of the atoms of certain elements. But what was the cause of radioactivity? What was so special about those elements? And what was the radiation made of? It was the great experimental physicist Ernest Rutherford who supplied the answers to these questions.

In 1899, Rutherford took a sample of uranium and wrapped it in layers of thin aluminium foils. As more foil was added, the effect was to reduce the amount of emitted radiation which could penetrate the total amount of foil. The experiment revealed that there were two types of radiation being emitted from the uranium sample with greatly different abilities to penetrate the foil. Rutherford called these two types of radiation *alpha* and *beta* (named after the first two letters of the Greek alphabet).[3] Alpha rays could be quite easily stopped by the foil, whereas the beta rays were more penetrating. Also, it was observed that the two types of rays were deflected by a magnetic field, suggesting that the alpha rays were composed of positively-charged particles, while the beta rays were composed of negatively-charged particles. Also, the alpha rays were deflected by a smaller amount than the beta rays, suggesting that the alpha particles had much greater mass than the beta particles.

---

[3] A third type of radiation called *gamma* radiation was discovered later. Gamma radiation is a form of high-energy electromagnetic radiation (such as light or X-rays). We will not be considering gamma rays further in this book.

What particles do we know which have negative charge and low mass and so have the same properties as the beta particles? Yes, electrons have negative charge and low mass. Indeed, further experiments revealed that the beta particles were, in fact, just electrons being emitted from the atoms at extremely high speed (high energy).

But what were the alpha particles made of? In 1909, Ernest Rutherford found the answer.

Rutherford was famous for creating ingenious experiments using the very limited apparatus available at that time. As an example, to examine alpha particles Rutherford placed a sample of radioactive radon gas in a thin-walled glass tube and then surrounded that tube with a second glass tube which had a thicker wall. The alpha particles emitted by the radon were able to penetrate the first glass wall but could not escape the second wall. As a result, the alpha particles became trapped in the space between the two glass walls. After a week, the substance between the two walls was analysed and found to be helium, which meant that alpha particles are composed of the nuclei of helium atoms.

The nucleus of a helium atom is composed of two protons and two neutrons. Rutherford's discovery, therefore, explained why alpha particles had positive electric charge (supplied by the two protons in the helium nucleus), and also had greater mass than the beta particles (which were just electrons).

The alchemists

The type of element (hydrogen, helium, uranium, etc.) is defined by the number of protons it has in its atomic nucleus. If alpha particles (which each contain two protons) were being emitted from atoms, then that implied that the element was changing into a different element, the atoms of which had two fewer protons. To change an element into a different element (for example, changing lead into gold) had long been the dream of alchemists. However, until the start of the 20th century it was considered impossible for any chemical element to change into a different chemical element – a process known as *transmutation*.

However, all that changed when Rutherford – together with Frederick Soddy – discovered that a sample of the radioactive element thorium appeared to be converting into radium. When Soddy realised what was happening, he shouted out "Rutherford! This is transmutation!", to which Rutherford replied "For Christ's sake, Soddy, don't call it transmutation. They'll have our heads off as alchemists."

If an element is capable of spontaneously changing into a different element, it would be very desirable to find a way of quantifying that change. For example, if I have a sample of uranium, I would like to know how long it would take for that sample to change into a different element. In order to be able to quantify the speed of the transmutation, it would appear we would need to know the precise details underlying radioactive decay (in physics, whenever an unstable particle transforms into a different particle it is called "decay"). However, the physics which underlies radioactive decay is quantum physics, and – if you have read my fourth book – you will know that all quantum processes are fundamentally random. In other words, you can never analyse those processes to a deeper level to discover how they work – you just have to accept that they are completely random. So that leaves us with a problem: if completely random quantum physics underlies radioactive decay, then how can we ever hope to develop a precise formula which describes the transmutation?

Well, even though we can never accurately predict when the decay of a single atom will occur, if we have billions of decay events then we can determine the average likelihood of a decay occurring and we can use that to make reliable predictions (the same method a casino uses to be certain of making money from probabilistic events).

A single atom of uranium, for example, might decay in the next second, or it might take billions of years before it decays – we have no way of knowing, it is fundamentally random. However, if we are considering a sample of billions of atoms then we can use an averaging statistical measure to describe the rate of decay. The statistical measure which is generally used is called the *half-life* of an element. The half-life is defined as the length of time it would take for half of the atoms in a sample to decay.

As an example, the half-life of uranium is 4.5 billion years. That means if I have a sample of uranium and wait for 4.5 billion years, then half of my sample will have turned into a different element, leaving me with a sample containing only half the original amount of uranium. If I then wait another further 4.5 billion years, half of that remaining amount of uranium will turn into a different element, leaving me with only a quarter of the original amount of uranium.

Nuclear reactions

This discussion of radioactive decay has provided us with a clearer understanding of the cause of radioactivity. Radioactivity occurs in heavy elements whose nuclei are fundamentally unstable. Those elements are heavy precisely because their atoms have nuclei made of a high number of protons and neutrons. As Heinz Haber says in *Our Friend The Atom*: "Here Nature has crowded so many protons and neutrons into the nucleus that it becomes unstable". Parts of those heavy nuclei get radiated-away as those elements decay into elements which have more stable nuclei. It is the emitted particles which form alpha and beta radiation.

Let us now consider the effect of the two types of radiation. We will need to define two important numbers which describe an atomic nucleus.

The *atomic number* of a chemical element is the number of protons in the nucleus of each atom of that element. The atomic number is crucial because an element is uniquely defined by the number of protons in the nuclei of its atoms. For example, all hydrogen atoms contain one proton, and all oxygen atoms contain eight protons. Hence, the atomic number defines which element an atom represents.

The other important number which describes the composition of an atomic nucleus is the *mass number*. The mass number is the **total sum** of the number of protons and neutrons in the nucleus.

The conventional notation is to describe a sample of an element by writing its chemical symbol preceded by its atomic number (as a subscript) and its mass number (as a superscript). This convention is shown in the following example for uranium (the chemical symbol for uranium is "U"):

$$^{238}_{92}U$$

Mass number

Atomic number

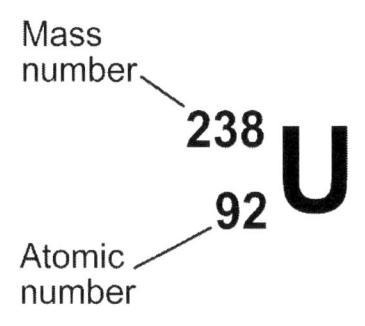

The previous example shows natural uranium, in which the nucleus has 92 protons and 146 neutrons, giving a total mass number of 238 (as shown in the previous image). Crucially, though, it is possible for an atom of a particular element to have the same number of protons **but a different number of neutrons,** in which case the substance is called an *isotope*. As isotopes of elements have different numbers of neutrons they therefore have different values for their mass number. The natural form of uranium shown in the previous diagram is uranium-238 where the "238" represents the mass number. Later in this book we will be examining the vital importance of the isotope uranium-235 for nuclear reactions.

Let us now consider what happens during the radioactive decay of an element. Remember, in alpha radioactive decay the nucleus emits a helium nucleus which is composed of two protons and two neutrons. So the emission of an alpha particle would decrease the atomic number by two (as the nucleus has lost two protons) and decrease the mass number by four (as the nucleus has lost a total of two protons and two neutrons). For the example of uranium-238, this would transmute the uranium atom into an atom of thorium:

$$^{238}_{92}\text{U} \xrightarrow{\text{Alpha decay}} {}^{234}_{90}\text{Th} + {}^{4}_{2}\text{He}\ \text{(Alpha particle)}$$

You can see from the previous diagram that thorium has an atomic number of 90 and a mass number of 234. You can also see the emitted helium nucleus forming the alpha particle.

Any process such as this in which atomic nuclei are modified is called a *nuclear reaction* (as opposed to a chemical reaction). When we describe a nuclear reaction using this notation, we must check two rules which are always true in any nuclear reaction:

1. The total number of protons and neutrons **before** the reaction is always equal to the total number of protons and neutrons **after** the reaction. In other words, the total mass number is always unchanged.

2. Electric charge is always conserved through a nuclear reaction.

Let us now examine how those two rules apply in the case of the alpha decay of uranium we have just considered.

Considering the first rule, and referring back to the previous diagram of alpha decay, you will note that the total number of protons and neutrons before the reaction (238, the mass number of uranium) is equal to the total number of protons and neutrons after the reaction (234 for the thorium nucleus, plus the mass number of four for the helium nucleus gives a total of 238). So the first rule is obeyed.

Now let us consider the second rule about conservation of electric charge. Every proton has a single unit of positive electric charge, so the total electric charge of the uranium nucleus before the reaction is 92 (the atomic number of uranium). After the reaction, the thorium nucleus has 90 units of positive electric charge (its atomic number), and the emitted alpha particle has two units of positive electric charge (the atomic number of helium). So electric charge is conserved through the reaction (92 = 90+2), and the second rule is therefore also obeyed.

The power of the atom

Alpha particles have enormous energy as they are ejected at about 5% of the speed of light. Plus, these particles are being constantly emitted as the individual atoms of a piece of radioactive material decays. Each click of a Geiger counter reveals the detection of one of these charged particles. You can see the use of a Geiger counter in the following direct link to *Our Friend The Atom*, immediately followed by the use of a cloud chamber to reveal the "constant shower" of radioactive particles emitted by a sample of radium:

http://tinyurl.com/radioactiveparticles

Now imagine the enormous quantity of energy which would be released if all those clicks of the Geiger counter happened at the same time. Or, equivalently, imagine if all those radioactive particles were emitted by the radium at the same time. Well, that is basically what happens in a nuclear explosion, with all the energy released in a few microseconds. According to Ernest Rutherford, a single gram of radium emits enough energy in its lifetime to raise a 500 ton weight a mile high.

Soddy commented on this energy release in a 1903 paper:

> *This reaction sets free more energy for a given weight than any other chemical change known. The energy of radioactive change must therefore be at least twenty-thousand times, and maybe a million times, as great as the energy of any molecular change.*

This gives us our first insight into why nuclear reactions (and explosions) are so powerful. For example, the detonation of dynamite releases about ten electron-volts of energy per molecule, whereas a nuclear reaction can release millions of electron-volts of energy per nucleus. Multiply that value by the trillions of atoms in a sample of nuclear explosive and you can start to understand the source of its extraordinary power. As Bruce Cameron Reed says in his book *The History and Science of the Manhattan Project*:

> *This begins to give you a hint as to the compelling power of nuclear weapons. An "ordinary" bomb that contains 1,000 pounds of chemical explosive could be replaced with a nuclear bomb that utilizes only 1/100 of a pound of a nuclear explosive. Thousands of tons of conventional explosive can be replaced with a few tens of kilograms of nuclear explosive.*

In a lecture in 1904, Soddy suggested presciently:

> *It is possible that all heavy matter possesses – latent and bound up with the structure of the atom – a similar quantity of energy to that possessed by radium. If it could be tapped and controlled what an agent it would be in shaping the world's destiny! The man who put his hand on the lever by which a parsimonious Nature regulates so jealously the output of this store of energy would possess a weapon by which he could destroy the Earth if he chose.*

Beta radiation

We have just examined alpha radiation so we will now examine beta radiation.

As explained earlier, Rutherford had discovered that beta radiation from an atom was composed of an emitted electron. This happens when a neutron in the nucleus converts into a proton – by a process which we will examine later in this chapter.

Now let us consider the effect of beta radiation on atomic number and mass number. The conversion of the neutron into a proton increases the atomic number by one (one more proton in the nucleus) while leaving the mass number unchanged (total number of protons and neutrons is unchanged). Hence, the following diagram shows how an atom of carbon-14 is transmuted into an atom of nitrogen-14 after beta radioactive decay (atomic number increases by one, mass number stays the same):

$$^{14}_{6}C \xrightarrow{\text{Beta decay}} {}^{14}_{7}N + e + \overline{v}_e$$

(Electron, Anti-neutrino)

You will see that an electron and a particle called an *antineutrino* are also emitted. We shall consider the antineutrino later in this chapter, but let us now consider why this conversion of a neutron in a proton necessarily results in an electron (the beta radiation) being emitted. Well, to see why, remember our two golden rules of nuclear reactions. The second golden rule says that electric charge must always be conserved through a nuclear reaction. With this in mind, remember a proton has positive electric charge, whereas a neutron has zero electric charge (an antineutrino also has zero electric charge). So when a neutron converts into a proton during beta decay we have a situation in which the amount of positive charge is increasing by one unit – which clearly breaks the law of conservation of charge. In order for balance to be restored, a negatively-charged electron must also be produced in addition to the proton. That electron represents the emitted beta radiation. Hence, the beta radiation emerges as a necessary result of the law of conservation of electric charge.

In 1946, a Bedouin shepherd named Muhammed edh-Dhib was herding his sheep through the barren rocky wilderness on the west side of the Dead Sea. High in the sheer cliffs above, he saw a small cave and decided to climb up to investigate. Deep in the cave he found a number of small pottery jars, and in those jars he found seven ancient parchment scrolls on which there was Hebrew writing.

Muhammed edh-Dhib had just discovered the Dead Sea Scrolls.

After selling the scrolls to an antiquities dealer, they caught the attention of an archaeologist who believed they dated back to Biblical times. But, unfortunately, there was no way of accurately dating the scrolls. It was left to science to provide the answer using a new technology which had just been developed.

It has just been described how the radioactive isotope carbon-14 will decay to nitrogen-14 via beta radiation. There is always a small amount of carbon-14 in the atmosphere, and this combines with oxygen to make carbon dioxide. Carbon dioxide is absorbed by plants during photosynthesis, and those plants are eaten by animals and humans, or that plant material is used by humans for creating parchment manuscripts. In other words, a significant amount of carbon-14 ends up in all organic material.

When a plant dies, it stops absorbing new carbon-14 from the atmosphere. However, the carbon-14 it absorbed during its life continues to decay. The half-life of carbon-14 is known to be 5,730 years. Therefore, a sample which was 5,730 years old would only have half the amount of carbon-14 compared to a living organism. After 11,460 years it would only have 25% of the amount of carbon-14 compared to a living organism. On that basis, in the late 1940s the science of *radiocarbon dating* was developed which was based on analysing the proportion of carbon-14 in dead organic material.

The radiocarbon dating system was calibrated by measuring the amount of carbon-14 in dead bristlecone pines which are found along the west coast of North America. Bristlecone pines were known to be the oldest living life-forms on Earth. The age of bristlecone pines could be measured accurately by counting the rings of bark which make up the tree trunk. Using this method, the oldest bristlecone pine was found to be 5,000 years old.

By measuring the amount of carbon-14 in an ancient bristlecone pine, a graph showing variation of carbon-14 with age was produced. The task now was to apply this new technique to dating the Dead Sea Scrolls. A sample of the linen wrapping of the scrolls was tested using radiocarbon dating in 1955 and was found to be 1,917 years old, with a plus or minus error of 200 years. The announcement of this first major success of radiocarbon dating had a major impact on archaeology, and radiocarbon dating laboratories have been created around the world.

Radiocarbon dating represents another way in which radioactivity has provided benefits to humanity. Remember that the basis of all radiocarbon dating is beta radioactive decay.

The weak interaction

Let us now consider beta radioactive decay in more depth. This section is quite technical and is not essential knowledge for atomic bomb construction, so you can skip to the next chapter if you wish. However, if you want to understand the origin of the antineutrino particle which is emitted during beta decay, here is the explanation.

It is now known that the protons and neutrons which make up the nucleus of an atom are not truly elementary particles. Instead, protons and neutrons are each composed of three particles called *quarks*, and it is these quarks which are considered to be truly elementary. Beta radiation is produced from the interactions between those quarks.

Protons and neutrons are composed of two different types of quarks called *up* quarks and *down* quarks. A proton is composed of two up quarks and one down quark. A neutron is composed of one up quark and two down quarks:

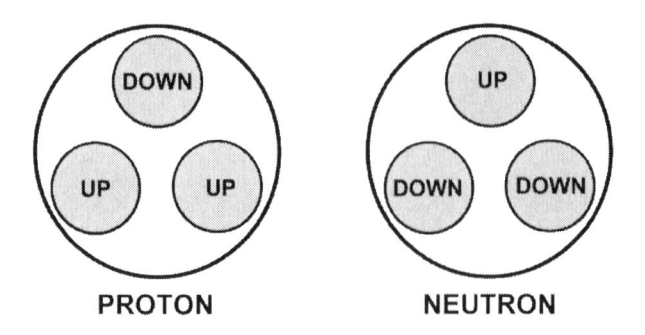

There is a symmetry between up and down quarks, which means it is possible for a down quark to turn into an up quark, and vice versa. This conversion is achieved via the *weak interaction* in which the quark interacts with a *boson* (a force-carrying particle). The following diagram is a simple *Feynman diagram* which shows how a down quark can change to an up quark when it interacts with a W boson, one of the bosons associated with the weak interaction. You can see that the W boson also modifies the path of the original particle – as if acted on by a force – so it is said that the W boson is the particle which carries the *weak force*:

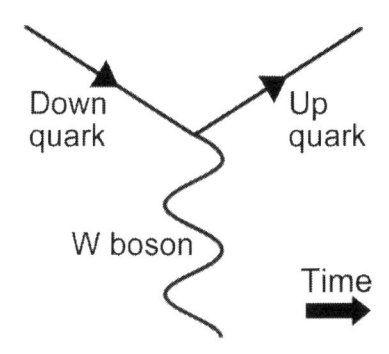

<figure>
Down quark → Up quark
W boson
Time →
</figure>

As stated earlier, a neutron is composed of two down quarks and one up quark. So this conversion of a down quark into an up quark via the weak interaction would have the effect of converting a neutron into a proton, which is what we know happens during beta radioactive decay. So we can now see why the weak interaction is so important for the emission of beta radiation.

For the next step in our analysis of beta radiation, we need to introduce a strange and elusive new particle: the *neutrino*. A neutrino is the only electrically-neutral matter particle, and it is virtually massless (i.e., it is extremely light) so therefore travels at speeds close to the speed of light. Because the neutrino is so light and has a ghostly nature, it passes through materials as though they were not there – including the materials which are used to make particle detectors. Unfortunately, a particle which flies straight through a particle detector without interacting with that detector is a particle which is not going to be detected easily. When Wolfgang Pauli predicted the existence of the neutrino in 1930 he said: "I have done a terrible thing. I have postulated a particle that cannot be detected!"

Don't get a neutrino and a neutron mixed up – their names might sound similar but they are completely different particles. As explained earlier, a neutron is composed of three quarks so it is not a genuinely elementary particle, whereas the neutrino is believed to be truly elementary (i.e., not composed of any other particles).

We have just seen that there is a symmetry between a down quark and an up quark, which is moderated by the weak interaction. There is a similar symmetry between a neutrino and an electron. The following diagram shows how a neutrino can change to an electron when it interacts with a W boson:

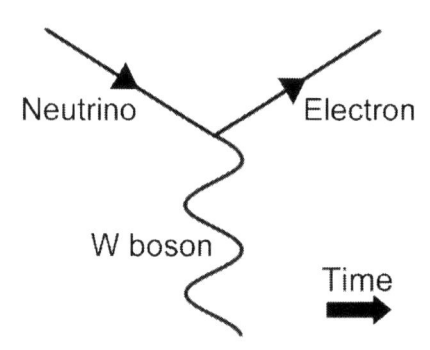

We have now generated two simple Feynman diagrams showing how the weak interaction can act to change down quarks into up quarks (and vice versa), and neutrinos into electrons (and vice versa). These simple Feynman diagrams which include just one vertex (junction) are called a *minimal interaction vertex*. These simple diagrams can be used to completely characterise the behaviour of a force.

FAINT FAIRY LIGHTS

Let us draw our two minimal interaction vertices again, and, so we can refer to them, let us number them as diagram 1 and diagram 2:

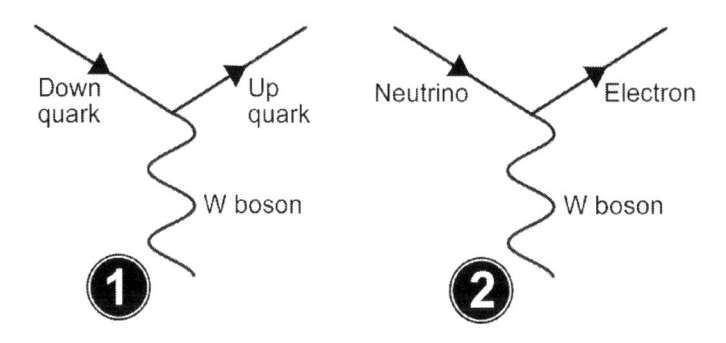

Diagram 1: Down quark → Up quark, emitting W boson.
Diagram 2: Neutrino → Electron, emitting W boson.

These are the two diagrams which interest us as these are the diagrams which are relevant to beta radioactive decay. Let us now use these two Feynman diagrams to construct the complete Feynman diagram of radioactive beta decay. This can be achieved because Feynman diagrams have a very useful feature: they can be rotated and the interaction they describe will still remain valid. On that basis, the following diagram includes both diagram 1 and diagram 2 which have been joined together. Both diagrams have been surrounded by grey dashed circles to make their position clearer. You will see that diagram 1 has been rotated slightly anti-clockwise, while diagram 2 has been rotated clockwise by a larger amount:

Neutron **Proton**

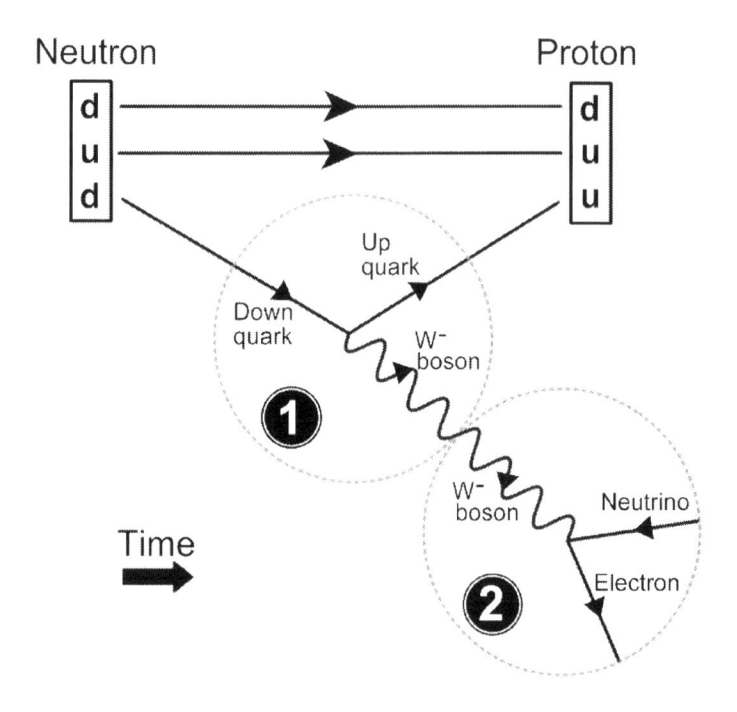

d → d
u → u
d   u

Down quark
Up quark
W⁻ boson
①
W⁻ boson
Neutrino
Electron
②

Time →

In this way, you can see how the complete diagram for beta radioactive decay can be constructed via these minimal interaction vertices – just like Lego!

As you can see from the previous diagram, the end result is that a neutron changes into a proton, and in the process it emits an electron and a neutrino (these two particles shown at the bottom right of the diagram). The emitted electron represents the beta radiation which is detected (remember, beta radiation is just a stream of electrons). The neutrino, meanwhile, does not interact with anything – it just shoots away at close to the speed of light straight through the walls of the laboratory as if they were not there and it is halfway to the moon in a matter of seconds.

Importantly, note that the arrow on the emitted neutrino seems to indicate that it is travelling backward in the time direction (you will see the forward time direction indicated by the arrow on the bottom left of the diagram). In a Feynman diagram, any particle which is travelling backward in time is equivalent to an *antiparticle* which is travelling forward in time. In other words, it is a particle of *antimatter* (see my fifth book for details). Hence, the emitted neutrino is not actually a neutrino at all: it is the antimatter equivalent of a neutrino which is called an *antineutrino*. This explains the origin of the antineutrino which is emitted during beta decay.

3

SPLITTING THE ATOM

As we have just seen, radioactive particles have high energy and are emitted from the nucleus at great speed. The question then arises: what would happen if one of these particles was directed to collide with a second nucleus? What would be the outcome?

That was the question first considered by Ernest Rutherford in 1909. Rutherford knew that the alpha particle had more mass than the beta particle, so an alpha particle colliding with a second atomic nucleus could potentially reveal the structure of that nucleus. With this in mind, Rutherford directed a steam of alpha particles at thin gold foil and placed a Geiger counter on the other side of the foil to detect the alpha particles as they passed through the foil. The intention was to consider the scattering of the alpha particles due to the nuclei in the gold foil.

To Rutherford's great surprise, some of the alpha particles bounced back from the foil. Rutherford realised this had to mean that there was a minute, comparatively-heavy, positively-charged nucleus at the heart of the atom. This was to be Rutherford's greatest discovery.

The following image shows Rutherford's laboratory at Cambridge, which was fairly typical of nuclear laboratories at that time. It looks very different from today's LHC, but no particle physics laboratory has made more historic discoveries than Rutherford's lab:

So the value of alpha particles as a nuclear probe was clear – but there was a problem. As described earlier, an alpha particle is a helium nucleus with two protons and therefore has positive electric charge. But a nucleus also has positive electric charge. So if an alpha particle is directed towards a second nucleus, the alpha particle will be repelled by the electric charge of that nucleus (two charges of the same sign will repel). This makes it very difficult to use a charged particle as a nuclear probe. The problem becomes more acute when an attempt is made to probe the nuclei of heavier elements as heavier elements have more protons in their nuclei and therefore have greater positive electric charge. As Richard Rhodes says in his book *The Making of the Atomic Bomb*: "Farther along the periodic table a barricade loomed. The naturally radioactive sources Rutherford used emitted relatively slow-moving alpha particles that lacked the power to penetrate past the increasingly formidable electrical barriers of heavier nuclei. For a time, the newborn science of nuclear physics stalled."

However, everything changed in 1932.

James Chadwick was a researcher at Rutherford's Cavendish Laboratory in Cambridge University. In 1932, in a letter to the British journal *Nature*, Chadwick announced the discovery of the neutron. As was described in Chapter One of this book, the atomic nucleus is composed of protons and neutrons. The neutron has approximately the same mass as the proton but it has no electric charge. It was realised very quickly that this meant that the neutron had the potential to be perfect for probing a nucleus as – unlike the positively-charged proton – it would not be repelled by the positive electric charge of the nucleus.

The neutron was to revolutionise the examination of the nucleus. The Nobel prize-winning nuclear physicist Hans Bethe once said that he considered everything before 1932 to be "the prehistory of nuclear physics, and from 1932 onward (after the discovery of the neutron) modern nuclear physics was born."

As Bruce Cameron Reed says in his book *The History and Science of the Manhattan Project*: "Neutrons would prove to be the gateway to reactors and bombs."

Artificial radioactivity

If nuclear physics had been stalled for over ten years since 1919, with the discovery of the neutron in 1932 it suddenly sprang into life. Progress would now be incredibly rapid as we shall see, with the discovery of nuclear fission in 1938, the construction of the first nuclear reactor in 1942, and the first man-made nuclear explosion in 1945 – events which all depended on the discovery of the neutron. After the discovery of the neutron, the race was on.

The first experimentalist to use neutrons to bombard atomic nuclei was Enrico Fermi. Fermi was a talented physicist based at the University of Rome who had been appointed to a full professorship at the age of 26. Fermi gained a reputation for having uncanny intuition about physics. It was even suggested that "Fermi had an inside track to God". His near infallibility at predicting the results of experiments meant his colleagues gave him the nickname "the Pope".

Fermi started his work on neutron bombardment in 1934, with a particular interest in bombarding the heavy metals which had many protons and neutrons in their nuclei. By bombarding various target elements, Fermi found he could make elements radioactive which would never normally be radioactive.[4] As an example, by bombarding gold with neutrons he found that a neutron was sometimes absorbed by the gold nucleus, creating a heavier nucleus with one more neutron than the natural form. However, the resultant isotope with its extra neutron was highly-unstable, and – as explained earlier – an unstable atom is radioactive. So the resultant isotope atoms were highly radioactive with a very short half-life of sometimes only a few minutes. As a result, Fermi would have to run down the corridor with his unstable radioactive isotopes to test them with a Geiger counter in a second room before the radioactivity faded away (due to the extremely short half-life).

---

[4] This echoed a discovery in 1932 by Marie Curie's daughter Irène and her husband Frédéric Joliot who had considered alpha particle bombardment of light elements.

SPLITTING THE ATOM

Let us consider the neutron bombardment of gold in more detail. The first stage is for the neutron to be absorbed by a gold nucleus, adding one to the mass number, and thereby creating an isotope of gold:

Neutron   **Gold**   **Gold isotope**

$$_{0}^{1}n + _{79}^{197}Au \rightarrow _{79}^{198}Au$$

(You might want to check that our two rules of nuclear reactions apply here: the neutron is electrically-neutral so electric charge is conserved before and after, and the total number of neutrons and protons before the reaction is equal to the total number of neutrons and protons after the reaction: 1+197=198).

But the additional neutron makes the nucleus unstable, so let us now concentrate on the gold isotope and consider the second stage of the process.

The nucleus of the unstable gold isotope decays via the beta radiation we considered in the previous chapter. That means a neutron in the nucleus transforms into a proton, with the emission of an electron (beta radiation):

Gold isotope   **Beta decay**   **Mercury**   **Electron (beta particle)**

$$_{79}^{198}Au \xrightarrow{\text{Beta decay}} _{80}^{198}Hg + e$$

So, as explained in the previous chapter, you can see that the result of the beta decay is to increase the atomic number of the gold nucleus by one, transmuting it into mercury ("Hg" is the chemical symbol for mercury). In the previous diagram you can see that the atomic number has increased from 79 (gold) to 80 (mercury).

Crucially, this shows that neutron bombardment can be used to increase the atomic number of the bombarded element, adding a proton to the nucleus and thereby transmuting it into a heavier element. We shall see later that this is the process by which heavy elements for nuclear weapons can be made. Indeed, as we shall see in Chapter Seven, the first atomic bomb used plutonium produced in an atomic reactor by this method of neutron bombardment.

If he could transmute gold into a heavier element via neutron bombardment, Fermi reasoned that he could also transmute uranium into a heavier element. However, that represented a much bigger deal: uranium is the heaviest natural element. If he was going to create a heavier element (a so-called *transuranic* element) then Fermi would be creating a substance which had never before existed on Earth.

But that appeared to be precisely what happened when Fermi bombarded uranium with neutrons. Beta emissions from the bombarded uranium appeared to suggest the usual increase in the atomic number by one. In June 1934, Fermi announced in *Nature* that his results "suggested the possibility that the atomic number of the element may be greater than 92 (the atomic number of uranium)." This announcement generated a great deal of excitement in Italy at the time, with some Italian journalists suggesting that this new element should be called "Mussolinium". However, this was to be one of the rare occasions when Fermi's intuition failed him.

In detecting a possible new heavy element, Fermi had been careful to eliminate the possibility that the radiation might be coming from some heavy element which was slightly lighter than uranium. However, he had never considered the apparently crazy possibility that it was coming from an element which was only half the weight of uranium, as if the uranium nucleus had been split in two. After all, such an effect had never been seen before.

But that was his big mistake …

SPLITTING THE ATOM

The discovery of fission

In 1912, the German Kaiser decided he wanted to support German science – and create a symbol of increasing German strength – by building a new scientific institute. The result was the Kaiser Wilhelm Institute in a suburb of Berlin. The architect of the institute decided to incorporate a spiked German army helmet on top of a round corner tower. The intention was undoubtedly to flatter the Kaiser but, as you can see at the top right of the following photograph, the end result was rather comical:

Among the first physicists to be based at the new institute were Otto Hahn and Lise Meitner. Hahn had established his reputation by previously working with Rutherford, and Meitner had published papers on alpha and beta radiation, gaining a reputation as an accomplished experimentalist. Einstein was later to call her the "German Marie Curie" (even though she was, in fact, Austrian).

By the 1930s, Hahn and Meitner had acquired enough experience to be acknowledged experts in the field of radioactivity (Richard Rhodes called Hahn "the ablest radiochemist in the world"). In 1935, they became intrigued by Fermi's results of bombarding uranium with neutrons and the claim that transuranic elements were being created. Hahn and Meitner decided to investigate Fermi's claim, and to help them in their task they enlisted the chemist Fritz Strassmann.

However, the investigations of Hahn and Meitner were to be derailed by the darkening atmosphere in Europe. In March 1938, Nazi Germany invaded Austria (the invasion was called the "Anschluss"). With Austria now part of the German Reich, Lise Meitner became a German citizen. This was a worrying development because Meitner was Jewish, so Germany's many anti-Semitic laws suddenly applied to her. Firstly, as part of the persecution of Jewish academics, Meitner's research funding was withdrawn. Meitner decided she had no choice but to flee Germany quickly. Niels Bohr arranged for her to find a research position at a physics institute in Sweden.

Hahn and Strassmann continued their research into the neutron bombardment of uranium, and kept Meitner informed of their progress. Hahn and Strassmann were discovering some very puzzling results. They appeared to be detecting barium as one of the side-products of the bombardment. This seemed to make no sense whatsoever as barium has an atomic number of 56 (56 protons in its nucleus), whereas uranium has an atomic number of 92. As was described earlier, alpha and beta decay might modify an atomic number by one or two places, but there was no known explanation for a modification of 36 places (92 minus 56).

However, further extensive checking proved that this seemingly inconceivable result was genuine: barium was being generated. Hahn and Strassmann were excited, but they were also completely stunned. How could a nucleus of approximately half the size of the uranium nucleus be produced? It was as though the uranium nucleus was being split in two!

Hahn contacted Meitner by mail, desperate to obtain a solution to the puzzle: "Perhaps you can suggest some fantastic explanation", he wrote. "We understand that it really **can't** break up into barium." At that moment, Meitner was being visited by her nephew, Otto Frisch, who was also a nuclear physicist who was working in Copenhagen with Niels Bohr. They went for a walk into the nearby Swedish forest to clear her heads and to try to imagine a possible solution.

At one point, they both sat down on a log to discuss the problem. Meitner realised that a possible solution might come from the "liquid drop" model of the atomic nucleus which had been suggested by Niels Bohr. According to Bohr, the protons in the nucleus were loosely held together by a form of surface tension in much the same way as a small drop of liquid. The result was that the nucleus was not a hard ball at all, but instead resembled a "wobbly drop" of liquid. In that case, the addition of just a single neutron to the nucleus might well be enough to make the whole wobbly nucleus unstable, and it might well split in two.

Meitner found a pencil and a scrap of paper in her purse and drew some rough shapes to explain her idea. You can see a version of Meitner's shapes in the following diagram. After the addition of the extra neutron, the unstable nucleus wobbles and becomes elongated. A narrow "waist" appears, giving the nucleus the shape of a dumbbell. Each end of the dumbbell would repel the other end due to the intense electrical repulsion between the protons. This would result in the two halves flying apart and forming two smaller nuclei:

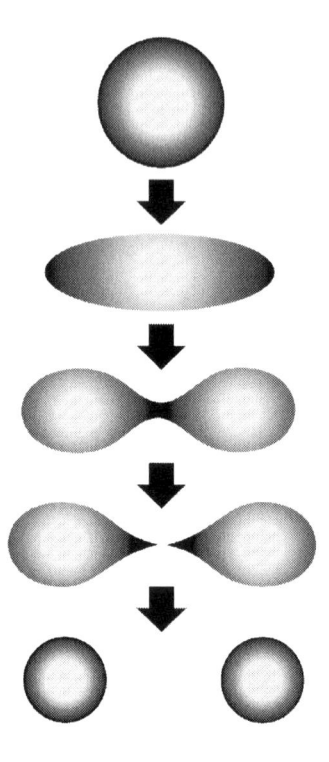

Otto Frisch coined the term *nuclear fission* to describe this splitting of a nucleus into two roughly equal parts. Let us consider this proposed fission reaction in more detail.

The proposal was that a nucleus of uranium-235 was absorbing a neutron, and that resulted in the nucleus splitting into a nucleus of barium and a nucleus of krypton, as described by the following reaction:

$${}_0^1 n + {}_{92}^{235}U \xrightarrow{\text{Fission}}$$

Neutron — Three neutrons

$${}_{56}^{141}Ba + {}_{36}^{92}Kr + 3[{}_0^1 n]$$

However, there was one other important factor to be taken into account. A nucleus of uranium-235 is so large that it needs 143 neutrons to act as "glue" to hold the nucleus together (235-92=143). This is more neutrons than is required in total by a nucleus of barium plus a nucleus of krypton. Hence, there would be an excess of neutrons after the reaction. You can see that the previous reaction also shows three neutrons being expelled. We will soon see that these three extra neutrons held the key to the all-important fission chain reaction.

You might like to check that the previous reaction with its three extra neutrons satisfies our golden rules for electric charge being conserved (92=56+36) and the total number of neutrons and protons being conserved (235+1=141+92+3).

Meitner and Frisch then turned their attention to calculating the amount of energy which would be released by this fission reaction, and the result left them totally astonished …

The nuclear mousetrap

Immediately after a nucleus fissions, two smaller nuclei are created, and those are positioned a microscopic distance apart. As both of those nuclei are heavily positively-charged, there is a huge electrical repulsion between them – they would actually fly apart at 3% of the speed of light. Meitner and Frisch calculated that this represented an energy of about 200 million electron-volts, which is enormous when you consider that – as stated in the previous chapter – the detonation of dynamite releases only about ten electron-volts per molecule. This meant that a nuclear explosion based on fission would be twenty million times more powerful than a conventional explosion using the same amount of material. One kilogram of uranium would have the same explosive power as twenty million kilograms (equivalently, twenty thousand tons) of conventional explosive.

As another way of looking at it, Otto Frisch calculated that the energy from each split uranium nucleus would be enough to make a grain of sand visibly jump. That might not sound very impressive, until you realise that in each gram of uranium there are 2.5×10^{21} atoms, which looks even more amazing when it is written as 2,500,000,000,000,000,000,000 atoms. It has been estimated that there are "only" 10^{18} grains of sand on all the beaches on Earth, which means a single gram of uranium possesses enough energy to make all the grains of sand on all the beaches on Earth jump – quite high. It is clear how an atomic fission bomb can release enough energy to cause catastrophic damage.

Lise Meitner wondered about the source of this tremendous energy, and then she remembered a lecture she had attended in 1909 given by Albert Einstein. In the lecture, Einstein had derived his famous formula $E=mc^2$ which described the equivalence between mass and energy, and explained how a small amount of mass could release a huge amount of energy. Meitner realised that if a uranium nucleus split into two smaller nuclei of barium and krypton then the total mass of the barium and krypton nuclei would be very slightly less than the mass of the original uranium nucleus. Some mass was being lost during fission. Meitner calculated that the amount of mass being lost was approximately equal to a fifth of the mass of a proton, and if that amount of mass was substituted into the $E=mc^2$ formula it could be calculated that the equivalent amount of energy was 200 million electron-volts, which was precisely the amount of energy being released by fission.

So it was clear that this was the source of the energy being released by fission: some of the mass of the uranium nucleus was being converted into energy. Let us consider this storage and release of energy in more detail.

Protons have positive electric charge, so they would naturally tend to repel each other via the electric force (like charges repel, opposite charges attract). It is therefore difficult to hold protons together in an atomic nucleus, and this must be achieved by a force which is stronger than the repelling electric force. That force is the *strong nuclear force*, the force which holds protons together in the nucleus. The strong nuclear force has a very short range compared to the electric force. Imagine the electric force acting like a spring trying to push protons apart. Then, in order to form a nucleus, protons must be pushed together against that electrical repulsion until the protons are very close together and then the short-range strong force can lock them into place like a latch.

The mechanism resembles a mousetrap. The electric force acts like the spring on the mousetrap. You have to put energy into the mousetrap in order to bend the spring and prime the trap, and then the strong force acts to lock the trap in place:

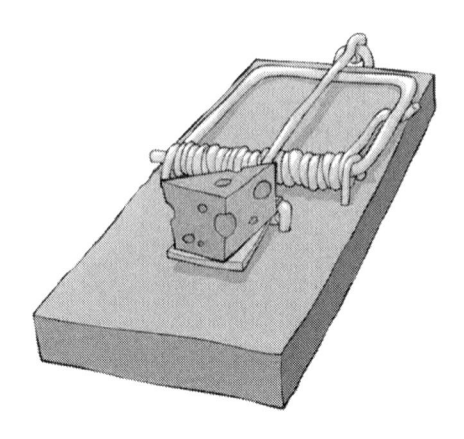

In the Disney television programme *Our Friend The Atom*, Heinz Haber compares the energy trapped in the nucleus to a mousetrap which has been set. Here is a direct link to the relevant point in the programme:

http://tinyurl.com/fissionmousetrap

How does the strong force work? How can it hold the protons together? Well, when protons and neutrons are pushed very close together, the strong force locks them into place by releasing an amount of energy via gamma radiation. This released energy is called the *binding energy*. If you then wanted to pull the neutrons and protons apart again to return the system to its initial state – fighting against the attractive strong force – you would have to do work to put energy into the system which is equal to that lost binding energy. This is due to the law of conservation of energy: if energy was lost, you will have to replace that energy to return the system to its original state. So it can be seen how the binding energy acts to lock protons and neutrons together via the strong force.

This release of binding energy during the formation of nuclei – by compressing protons and neutrons together – is called *nuclear fusion*. This process occurs in the centre of stars, forming light elements such as helium from hydrogen, with the release of huge amounts of energy.

But for heavy elements being fused together, with many protons in their nuclei, the electrical repulsion becomes an ever-larger factor. In other words, the "spring" on the mousetrap becomes much stronger, and the mousetrap becomes harder to set (it becomes harder to push the nuclei together due to the electrical repulsion). In that case, the energy required to push the protons together is larger than the binding energy that would be released when the strong force locks the protons together. Therefore, the generation of energy via nuclear fusion is then no longer possible. Any element heavier than iron cannot be used for energy production via nuclear fusion inside a star.

But when a star explodes in a massive supernova explosion, the pressures and temperatures are so great that even heavy elements, heavier than iron, can be fused together, pushing against the tremendous strength of the electrical "spring". This is how uranium, the heaviest naturally-occurring element is formed: all the uranium now found on Earth was created in one or more supernova explosions over six billion years ago.[5] Luckily, the 4.5 billion-year half-life of uranium explains why not all of the uranium on Earth has vanished due to radioactive decay over that period.

The energy then lies trapped in the uranium nucleus for billions of years, like a primed mousetrap. And when the nucleus fissions, the trap explodes into action. As we have

---

[5] *The Cosmic Origins of Uranium*, World Nuclear Association, http://tinyurl.com/uraniumorigins

seen, the two halves of the fissioned nucleus fly apart with tremendous energy because of the intense electrical repulsion between them. The energy stored in the mousetrap is finally released. As Robert Serber says in his book *The Los Alamos Primer*, the two nuclei "would fly apart with an amount of energy equal to the work that went into pushing them together."

So when we are watching a nuclear explosion, we are watching the energy which has been trapped in atoms by supernova explosions billions of years ago, energy which has waited half the age of the universe before being set free. Maybe that thought should fill us with increased awe when we watch the extraordinary phenomenon that is a nuclear explosion.

4

THE CHAIN REACTION

At this point in our story, we must introduce one of the most fascinating characters in the story of the atomic bomb. This is a man who played a seemingly peripheral role during the development of nuclear physics in the early decades of the 20th century, but he had a vision of the true potential of nuclear power and, perhaps more importantly, he was the first man to see the possibility of a nuclear chain reaction. As we shall soon see, he then went on to play a central role in the decision to develop nuclear weapons by enlisting the help of none other than Albert Einstein.

That man was Leo Szilard.

Leo Szilard (the "s" is silent) was born in Budapest in Hungary in 1898. After deciding he could not achieve his ambitions in Hungarian universities, he moved to Germany. He started a course on engineering, but switched to study physics at the University of Berlin because he heard that several Nobel laureates including Albert Einstein and Max Planck taught at the university. At this point we start to see Szilard as a man who was perhaps unsure of the direction he wanted to take in life, a man who for most of his life lived in rented rooms, living out of suitcases – which was perhaps not a bad idea given the political situation in Germany at the time (Szilard was a Jew): "All I had to do was turn the key of the suitcase and leave when things got too bad."

Although Szilard was studying physics, he was more of an inventor and entrepreneur than an academic. In fact, he might even be described as something of a "hustler", always with an eye for a money-making scheme. Rather bizarrely, he approached Albert Einstein with an idea for a new type of refrigerator which had no moving parts. Even more bizarrely, Einstein agreed to work with Szilard to develop the fridge, with Einstein using his earlier patent office experience to formulate a patent application.

Yes, the Einstein fridge really was a thing.

As we shall see later, Szilard's strong working relationship with Einstein was to play an important role in the development of the atomic bomb.

Perhaps Szilard's greatest talent was an almost uncanny ability to visualize the future and sense trends – perhaps a necessary talent for any successful entrepreneur. In the late 1920s he became aware of the promising advances in nuclear physics and was particularly interested to hear of Chadwick's discovery of the neutron in 1932. Szilard immediately realised the potential of the neutron as a way to probe the nucleus without being repelled. Might the neutron also provide a way to unlock atomic power?

In 1933, Adolf Hitler was appointed Chancellor of Germany, prompting Szilard to move to London in anticipation of the Nazi persecution of the Jews. It was in London that Szilard read a remarkable book by another visionary figure: the English science fiction writer H.G. Wells.

H.G. Wells had become aware of the discoveries of Rutherford and Soddy (which were described in Chapter Two) which suggested that there was a huge reserve of energy contained in the nucleus of an atom. Let us remind ourselves of Soddy's alarming prediction in 1904:

> *It is possible that all heavy matter possesses – latent and bound up with the structure of the atom – a similar quantity of energy to that possessed by radium. If it could be tapped and controlled what an agent it would be in shaping the world's destiny! The man who put his hand on the lever by which a parsimonious Nature regulates so jealously the output of this store of energy would possess a weapon by which he could destroy the Earth if he chose.*

Such a dramatic statement must have resonated with H.G. Wells who, in 1913, wrote a remarkably prophetic book called *The World Set Free*. In that book, Wells imagines a world in which nuclear power has been harnessed by mankind, but it has also been used to create atomic bombs. Szilard described the plot of *The World Set Free* in his own words:

> *Wells describes the liberation of atomic energy on a large scale for industrial purposes, the development of atomic bombs, and a world war which was apparently fought by an alliance of England, France, and perhaps including America, against Germany and Austria, the powers located in the central part of Europe. He places this war in the year 1956, and in this war the major cities of the world are all destroyed by atomic bombs.*

At the time, Szilard was living at the Imperial Hotel in Russell Square in London, not far from the British Museum, living off his savings of £1,500 (£100,000 in today's money). On the morning of September 12th 1933, Szilard was lounging in the hotel lobby when he read a report in *The Times* newspaper of one of Rutherford's recent speeches at the nearby meeting of the British Association for the Advancement of Science. Rutherford had explained how alpha particles could be accelerated to such energies that they might be able to release energies from atomic nuclei. But Rutherford said he believed this would require more energy input than could be gained. Rutherford said that "anyone who looked for a source of power in the transformation of the atoms was talking moonshine (meaning foolish talk, or nonsense)."

According to Richard Rhodes: "All of which made Szilard restive. The leading scientists in Great Britain were meeting and he wasn't there. He was safe, he had money in the bank, but he was only another anonymous Jewish refugee down and out in London, lingering over morning coffee in a hotel lobby, unemployed and unknown."

Szilard was annoyed by Rutherford's negativity. He threw down his newspaper and stormed out of the hotel onto the street. Walking down the road, Szilard then had a flash of inspiration which he described in his own words:

> *Pronouncements of experts to the effect that something cannot be done have always irritated me. That day as I was walking down Southampton Row and was stopped for a traffic light, I was pondering whether Lord Rutherford might not prove to be wrong. As the light changed to green and I crossed the street, it suddenly occurred to me that if we could find an element which is split by neutrons and which could emit two neutrons when it absorbed one neutron, such an element, if assembled in sufficiently large mass, could sustain a nuclear chain reaction, liberate energy on an industrial scale, and construct atomic bombs.*

According to Richard Rhodes: "Szilard was not the first to realize that the neutron might slip past the positive electrical barrier of the nucleus; that realization had come to other physicists as well. But he was the first to imagine a mechanism whereby more energy might be released in the

THE CHAIN REACTION

neutron's bombardment of the nucleus than the neutron itself supplied." In particular, this was the first time that anyone had conceived of the principle of using neutrons to generate a chain reaction.

So what did Szilard mean by a "nuclear chain reaction"? Well, as he suggested, if a nucleus can be split (fissioned) by a neutron, then that split nucleus can produce additional neutrons (as described in the previous chapter). Those neutrons can then proceed to fission other nuclei. If a split nucleus emits two neutrons, then that will result in two further nuclei being fissioned. If those two nuclei emit two neutrons each, then that means there will be four neutrons in total produced, and those neutrons can proceed to split a further four nuclei. By this constant doubling of numbers, the total number of fissioned nuclei will rise extremely rapidly to a huge number.

Remember the analogy from the previous chapter of the energy trapped in an atomic nucleus being like a mousetrap. In the Disney television programme *Our Friend The Atom*, Heinz Haber performs an excellent visual demonstration in which many mousetraps (representing nuclei) are used to generate a "chain reaction" of mousetraps and ping-pong balls. Here is a direct link to the relevant point in the programme:

http://tinyurl.com/mousetrapreaction

The programme segment also considers the discovery of fission by Hahn and Strassmann.

The mousetrap chain reaction idea was also used in a clever (and destructive) Pepsi advert:

http://tinyurl.com/mousetrapchainreaction

The following diagram shows the start of a chain reaction in which each split nucleus releases three neutrons:

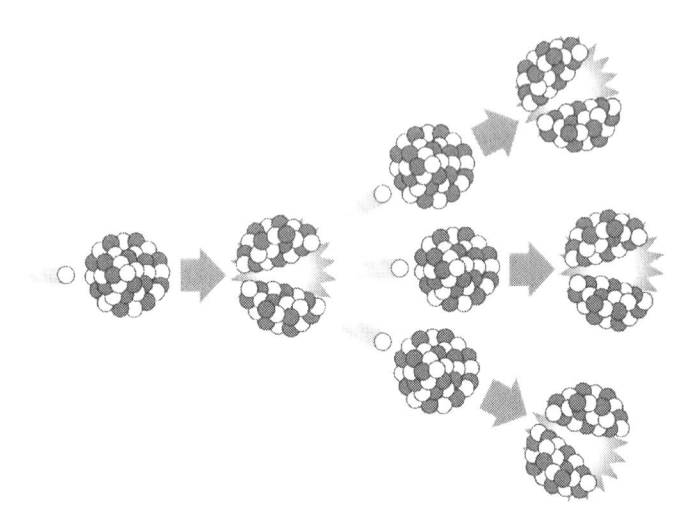

Bearing in mind that a neutron from a fissioned nucleus moves at about 5% of the speed of light, this chain reaction is clearly going to release its energy extremely quickly. With a great deal of energy released with each fissioned nucleus, Szilard realised that the total amount of energy released by an uncontrolled chain reaction could be used to create an atomic bomb.

But Szilard had his idea in 1933, and in the previous chapter we saw that nuclear fission was not discovered by Hahn and Strassmann until 1938. So, for the time being, Szilard's idea about a nuclear chain reaction would have to remain just an idea.

The New World

As another example of Leo Szilard's uncanny foresight, he saw the oncoming Second World War with great clarity. In 1936 he announced his intention to "stay in England until one year before the war, and then move to New York City". It was a very strange thing to say as no one in 1936 could be sure if there would be a war, or when it would start. But Szilard's prediction was perfectly accurate: he moved to America in 1938, precisely one year before the war began.

Szilard was joining a group of brilliant physicist refugees from Nazi Germany who were either Jewish or had Jewish family connections. The group included Enrico Fermi, Eugene Wigner, Hans Bethe, Edward Teller, John von Neumann, and Emilio Segrè. According to Emilio Segrè: "America looked like the land of the future, separated by an ocean from the misfortunes, follies, and crimes of Europe." All of those aforementioned refugee physicists were destined to work on the American project to develop the atomic bomb, designed to be deployed against their former persecutors in Germany.

In 1939, Leo Szilard travelled from New York to Princeton to visit his friend Eugene Wigner who informed Szilard of the recent sensational news of the discovery of fission by Hahn and Strassmann. According to Szilard:

> *Wigner told me of Hahn's discovery. Hahn found that uranium breaks into two parts when it absorbs a neutron. When I heard this I immediately saw that these fragments, being heavier than corresponds to their charge, must emit neutrons, and if enough neutrons are emitted then it should be, of course, possible to sustain a chain reaction. All the things which H.G. Wells predicted appeared suddenly real to me.*

Szilard's phrase describing the two resulting fissioned nuclei as being "heavier than corresponds to their charge" is interesting. This principle was described in the discussion of fission in the previous chapter. A nucleus of uranium is so large that it needs many neutrons to act as "glue" to hold the nucleus together. But this is more neutrons than is required in total by the two resulting fissioned nuclei, hence the phrase describing the uranium nucleus as being "heavier than corresponding to its charge". There would be more neutrons than required. This fact led to Szilard's suspicion that neutrons would be released by fission.

As Szilard realised earlier, in order for there to be a chain reaction, a uranium nucleus had to emit two neutrons when it was absorbed and split by one neutron. Whether or not this was the case was something Szilard had to determine.

Szilard was still living an unsettled life in America, but he arranged to use a laboratory in Columbia University in New York City for three months. In his experiment, he bombarded uranium with neutrons in order to fission the uranium nuclei. Any neutrons being released as a result of fission would be detected and displayed on an oscilloscope. According to Szilard: "If flashes of light appeared on the screen, that would mean that neutrons were emitted in the fission process of uranium and this in turn would mean that the large-scale liberation of atomic energy was just around the corner. We turned the switch and saw the flashes. We watched them for a little while and then we switched everything off and went home."

Analysing the results, Szilard found that there were approximately two neutrons released per fissioned nucleus, which would be enough to sustain a chain reaction. According to Szilard: "That night there was little doubt in my mind that the world was headed for grief."

The letter from Einstein

Leo Szilard was now confident that a chain reaction would soon be achieved, and there was now a race with Nazi Germany to create the first atomic bomb. However, initial contacts with the military were not promising. The only interest came from a naval physicist who worked on submarine propulsion and was attracted to the idea of a power source which did not require oxygen (remember the discussion of the USS Nautilus in Chapter One).

Szilard became frustrated at the lack of progress. According to Richard Rhodes, at this point: "despite his Olympian ego not even Leo Szilard felt capable of saving the world entirely alone." Szilard contacted his fellow refugee physicists Edward Teller and Eugene Wigner and asked for their help. Wigner, in particular, felt it was essential to inform the U.S. government, maybe even going straight to the president if necessary. Szilard wondered if he could get his old refrigerator engineer friend, Albert Einstein, to come on board. Einstein was no longer producing useful work, but he was still the most famous physicist in the world and his word would carry weight in Washington D.C.

Despite being unable to swim, Einstein was a keen sailor, and Szilard learned that Einstein was away sailing in Peconic Bay on Long Island (Einstein had a reputation as a lousy sailor, getting stuck on a sand bar once, and on another occasion almost drowning when his boat capsized). Einstein was staying at a summer house by the sea on Nassau Point.

Maybe physicists are generally not good at directions and vehicles, because Szilard could not drive and had to rely on his friend Eugene Wigner to drive to Long Island. They set out on their journey on Sunday, July 16th, 1939. Inevitably, they got lost and spent hours circling the narrow country lanes, trying in vain to locate Einstein's house.

On the verge of giving up, Szilard saw a seven-year-old boy standing on the curb, so Szilard asked him: "Do you know where Professor Einstein lives?" Luckily, the boy knew the way and was able to direct them to Einstein's house at the end of Old Cove Road. It is interesting to imagine how the world might have turned out differently if Szilard and Wigner had never met that seven-year-old boy, and therefore never managed to find Einstein.

Once Szilard explained the principle of the nuclear chain reaction to Einstein, Einstein exclaimed "I never thought of that!" Einstein agreed to dictate a letter which was to be delivered to Franklin Roosevelt, the president of the United States. Here is an edited portion of the letter which was sent:

```
                        Albert Einstein
                        Old Grove Rd.
                        Nassau Point
                        Peconic, Long Island

                        August 2nd, 1939

F.D. Roosevelt,
President of the United States,
White House
Washington, D.C.

Sir:
     Some recent work leads me to expect that
the element uranium may be turned into a new and
important source of energy in the immediate
future. I believe therefore that it is my duty
to bring to your attention the following facts
and recommendations:
     It may be possible to set up a nuclear
chain reaction in a large mass of uranium, by
which vast amounts of power and large quantities
of new radium-like elements would be generated.
```

```
Now it appears almost certain that this could be
achieved in the immediate future.
    This new phenomena would also lead to the
construction of bombs, and it is conceivable -
though much less certain - that extremely
powerful bombs of a new type may thus be
constructed. A single bomb of this type, carried
by boat and exploded in a port, might very well
destroy the whole port together with some of the
surrounding territory.
    In view of this situation you may think it
desirable to have some permanent contact
maintained between the administration and the
group of physicists working on chain reactions
in America.

                        Yours very truly,

                        Albert Einstein.
```

When Roosevelt was handed the letter and read it, he agreed that this required investigation, but there still no great urgency and only limited resources were assigned to the investigation.

However, everything changed on the morning of December 7th, 1941. On that morning, Japan attacked the U.S. Pacific Fleet at Pearl Harbor, triggering the entry of America into the war. From that point, the race to build the first atomic bomb had begun. Codenamed the *Manhattan Project*, it was to be the largest single industrial and scientific project in the history of the world.

In his book *Trinity*, Jonathan Fetter-Vorm describes the challenge of the Manhattan Project:

> *For all the money and infrastructure invested, the challenge of the Manhattan Project was still rather mystical. Somehow, a very small chunk of glowing metal was going to be transformed into the largest explosion that humans had ever made.*

Oppenheimer

Robert Oppenheimer had been a precocious child. He was born in 1904 to an ambitious family. When his grandfather saw his five-year-old grandson playing with bricks, he gave him an encyclopaedia of architecture. Oppenheimer developed an interest in science at an early age – mainly chemistry and mineralogy – and at the age of twelve he gave a lecture to the members of the New York Mineralogical Club. He had a wide range of other interests, including writing poems at the age of ten. At school, Oppenheimer did not hide the fact that he felt superior to the other children, telling them: "Ask me a question in Latin and I will answer you in Greek". He was described by a childhood friend as "very frail, very shy, very brilliant of course, and very superior."

Rather a sickly child, at the age of eighteen he was sent to spend the summer at a ranch in New Mexico to toughen him up before university. The ranch was in the mountains northeast of Santa Fe, and Oppenheimer thrived in the wilderness environment, chopping wood and learning to ride horses. The area was to make a lasting impression on Oppenheimer who was to say later that his two great loves were "physics and desert country".

At Harvard to study chemistry, Oppenheimer proved adept in a wide range of additional subjects including French literature and philosophy. He also wrote short stories and poetry, often giving an impression of being overly-sensitive and pretentious – and something of a drama queen. In a letter to a friend, he described his recent activities as: "Read Greek, committed faux pas, and wished I was dead. Voila".

In university, Oppenheimer extended his esoteric interests to include Hindu philosophy, while transferring his main field of study from chemistry to physics. However, Oppenheimer struggled with the laboratory work, "unable even to solder two copper wires together", and, as a result, he concentrated on theoretical physics. Oppenheimer went on to make significant contributions in the field of quantum mechanics.

Here is a photograph of Robert Oppenheimer:

In 1942, in a surprising move, Oppenheimer was selected to be the scientific director of the Manhattan Project. Oppenheimer was a surprising choice in many ways. Firstly, he was a theoretical physicist, rather than an experimental physicist. Oppenheimer may have been able to speak six languages, but surely someone in charge of running the largest-ever research and development project should have been able to solder two copper wires together.

Secondly, while Oppenheimer had no Nobel Prize he would be in charge of several physicists who did have Nobel Prizes. It was not so much a question of whether Oppenheimer was their intellectual equal – he was undoubtedly brilliant – it was more a question of whether the other physicists would be happy working under him.

Thirdly, Oppenheimer was well-known for having left-leaning political sympathies. Though he was not a communist, many of his friends were. Would he be a security risk? That was to be a question which dogged Oppenheimer throughout his career.

But what was impressive about Oppenheimer – and the reason he was selected to be project director – was his ambition and determination to make the project a success. Oppenheimer saw this as his opportunity to prove to others what he knew all along: that he was the best.

Los Alamos

One of the first decisions to be made was where to locate the development laboratory for the Manhattan Project. As Bruce Cameron Reed says in his book *The History and Science of the Manhattan Project*, the selected location had to satisfy the following criteria: "A laboratory site would have to be isolated, relatively inaccessible, have a climate that would permit year-round construction and operations, be large enough to accommodate a testing area, and be sufficiently inland to be secure from enemy attack."

Oppenheimer recommended the "desert country" of New Mexico where he spent summers riding horses as a youth. Specifically, he knew there was a ranch school in Los Alamos which would be ideal. The army engineers were concerned about the narrow access road from Santa Fe and the poor water supply, but otherwise agreed with Oppenheimer that it would be ideal.

The geography of Los Alamos is striking and unusual. It is situated on the slopes of the 13-mile-wide extinct volcanic crater called the Valles Caldera (or Jemez Caldera). Over a million years, rainfall down the slopes has carved a series of steep canyons, leaving a flat-topped hill on either side of each canyon. A small flat-topped hill like this is called a *mesa* ("mesa" is Spanish for "table"), and Los Alamos was spread out over four mesas. The following photograph shows the present-day town of Los Alamos spread over the four mesas, with the canyons visible. The location of the original Los Alamos laboratory is indicated by an arrow:

Because of its location on top of the mesa, the Los Alamos laboratory was given the nickname "The Hill".

For the remainder of 1942, and into the first few months of 1943, Oppenheimer had the job of touring the physics departments of American universities to persuade the best available physicists to join the Manhattan Project. His task was not made easier by the need to maintain strict secrecy. Although he could not describe the work in detail, he could state that this was important work which would probably end the war to end all wars. Almost every physicist he approached decided to join the project.

Security within Los Alamos was strict. In order to reduce the possibility of security leaks, the entire Los Alamos site was fenced-off. As a result, the facility operated like a small town, with a school, a general store, a family doctor, a library, and even a movie theatre. With a whole town having to be built from scratch, conditions were primitive. There were no paved roads, and fresh food and water was scarce. Los Alamos resembled a frontier town of the old American West. A wife of one of the physicists described her role as being "akin to the pioneer women accompanying their husbands across uncharted plains westward."

Initially, the military insisted on strict security in the scientific section of Los Alamos, with a need-to-know restriction on the transfer of information between sections. However, Oppenheimer argued that scientific progress could only be made if there was freedom of speech and free exchange of ideas. Eventually, the military agreed to complete scientific openness within the camp, but it came at a cost: the technical area of Los Alamos had to be entirely surrounded by high barbed-wire fence. For the refugee physicists from Europe, this generated the uneasy feeling that they were in a concentration camp.

The following photograph shows the entrance to the technical area of Los Alamos in 1944. The visual similarity to a concentration camp is clear:

The *Los Alamos Primer*

Because of the strict security, new scientific personnel arriving at Los Alamos were completely unaware of the work they would be doing or the goal of the Manhattan Project. In order to bring them up to speed quickly, it was decided to hold a series of introductory lectures. The task of giving the lectures fell to Robert Serber who was a former postdoctoral student of Oppenheimer. Serber's lectures were recorded in a 24-page document which was called the *Los Alamos Primer*.

The document represents everything that was known in 1943 about how to make an atomic bomb. In which case, it is surprising that this top secret document was declassified in 1965. Once declassified, the document was released into the public domain.

Here is a link to the original 1943 Los Alamos Primer document:

http://tinyurl.com/losalamosprimer

It is basically a recipe for making an atomic bomb.

The document was available on the Los Alamos website until 2001. However, everything changed after the September 11th attacks of that year. Because of the possibility of terrorists obtaining information about atomic bomb manufacture, the document was removed from the website on that day. But, as is the nature of the internet, once the document was released the genie could never be put back in the bottle.

Using the Los Alamos Primer as our starting point, let us now consider the physics of how to make an atomic bomb, starting with the most important calculation …

5

THE CRITICAL MASS

In this chapter we will discover the crucial importance of the volume of the fissionable material which is used to make an atomic bomb. We shall see that there exists a certain size of material at which point the chain reaction will be unstoppable and an explosion becomes a certainty. This amount of material is called the *critical mass*.

The accurate calculation of the value of the critical mass is the most important calculation in the construction of a nuclear bomb. Indeed, when you consider the impact of nuclear weapons, this is surely one of the most important calculations in the history of science. Therefore, this chapter considers the calculation of the critical mass. The calculation is inevitably highly mathematical, so most of the mathematical details have been included in the Appendix at the end of this book. I do not believe the detailed calculation of critical mass presented in this book can be found anywhere on the internet.

The ratio of volume to surface area

The existence of a critical mass relies on the fact that the ratio of an object's volume to its surface area is greater for larger objects than for smaller objects. In order to understand this principle, consider the following diagram showing three cubes of different sizes:

$$\frac{\text{Total volume of cubes}}{\text{Area of square sides}} = \frac{1}{6} = 0.166$$

$$\frac{\text{Total volume of cubes}}{\text{Area of square sides}} = \frac{8}{24} = 0.333$$

$$\frac{\text{Total volume of cubes}}{\text{Area of square sides}} = \frac{27}{54} = 0.5$$

The first cube at the top of the diagram is the smallest. We can say that its volume is equal to one unit of volume (one small cube). If you count the sides of that small cube, you will find that it has six sides. So the total surface area of the cube is six units of area. This means that the all-important ratio of volume to surface area is then 0.166, as shown on the diagram.

THE CRITICAL MASS

The second cube in the diagram has a total volume of eight small cubes (you can count them), and a total surface area of 24 units. Hence, the ratio of volume to surface area has increased to 0.333, as shown on the diagram.

The final cube, shown at the bottom of the diagram, is the largest. You can count that it has a total volume of 27 cubes, and a total surface area of 54 units. Hence, the ratio of volume to surface area has increased to 0.5, as shown on the diagram.

So larger objects have a greater ratio of volume to surface area. Interestingly, this explains why – in general – larger animals are found in cold climates, e.g., polar bears, and smaller animals are found in hot climates, e.g., rodents (I guess we are supposed to ignore the elephant in the room). A small animal has a larger surface area (proportionately) and so can cool down better in a hot climate. Whereas a larger animal has a proportionately smaller surface area so loses less heat through its skin. This principle is called *Bergmann's rule*.

The following diagram illustrates Bergmann's rule by showing how larger moose are found in more northerly parts of Sweden:

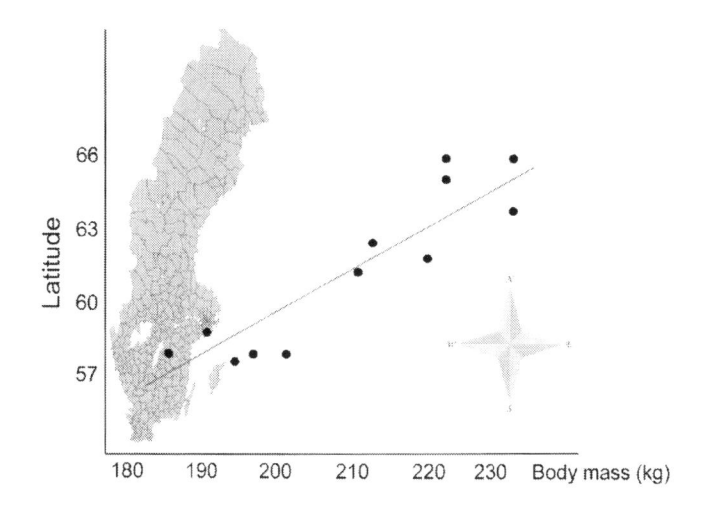

Why does this principle allow us to build a bomb based on nuclear fission? To put it simply, if we have a sufficiently large amount of fissionable material then the build-up of neutrons inside the volume of the material will be larger than the loss of neutrons through the surface area of the material. When that happens, the resultant chain reaction will increase exponentially and the result will be an explosion.

If we have a small amount of fissionable material then the surface area of that material will be proportionately quite large compared to its volume. As a result, the amount of neutrons being lost through the surface will be greater than the amount of neutrons being generated inside the material and the chain reaction will be unable to continue.

As the amount of mass is progressively increased, the surface area becomes proportionately smaller compared to the volume of the material, and so the loss of neutrons also becomes proportionately smaller. As the amount of mass continues to increase it is clear that a certain balance point will be reached at which point the loss of neutrons through the surface will be precisely equal to the build-up of neutrons inside the material. This amount of mass is called the critical mass. A sample of material this size – or larger – is going to explode.

This is an important point which we be using later in our calculation: if we have a volume of fissionable material, then **critical mass occurs when the build-up of neutrons inside the material is equal to the loss of neutrons out of the material.**

Put simply, the critical mass is the smallest amount of fissionable material which will cause a nuclear explosion. Therefore, determining the correct value of the critical mass is the most important task in designing a nuclear bomb.

Spherical coordinates

In our analysis of critical mass, we will need to use the most efficient and natural method of describing the distribution of neutrons within the material.

Normally, when we want to describe the shape of an object in three-dimensional space we would use Cartesian coordinates (as shown in the diagram below). We could then define any point in space by its (x, y, z) coordinates, where x, y, and z are the distances along the three axes:

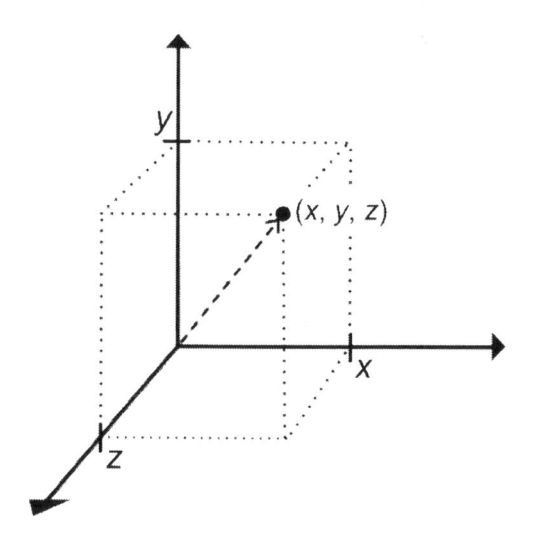

However, for describing the distribution of neutrons within a sample of fissionable material, we can use a better method of defining a point in space. This is because nuclear bombs almost always use a solid sphere of fissionable material as a critical mass (a solid sphere represents the most efficient use of the material). The most efficient means of defining a point in a solid sphere is to use *spherical coordinates*.

Spherical coordinates define a point in space using the distance from the centre of the space, and two angles of rotation, as shown in the following diagram:

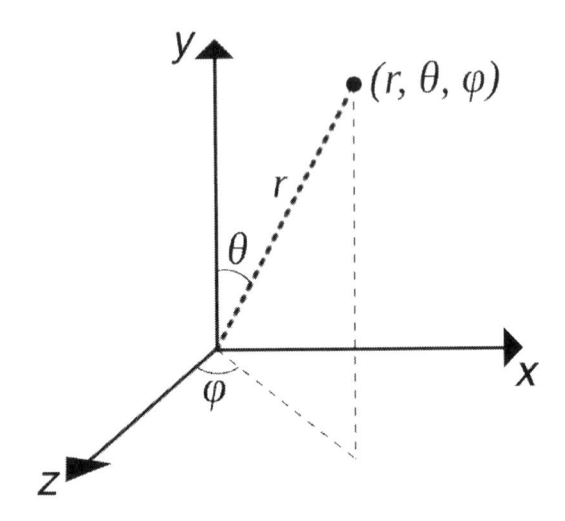

This method for defining a point in space is so efficient because our sphere of fissionable material has rotational symmetry: it does not matter how we rotate that sphere in space, it would still look the same. Because of that rotational symmetry, all we need to consider is the distribution of neutrons along the radial thick dashed line coming out of the centre of the material, as shown in the previous diagram. We do not need to be concerned about the two rotational angles because it does not matter how we rotate the line the distribution of neutrons along it will still be the same.

So, in this calculation of critical mass, we will only be interested in the distribution of neutrons in the one dimension along the line of length r coming from the centre of the material.

THE CRITICAL MASS

The build-up of neutrons

In this section we will consider the build-up of neutrons within a volume of the fissionable material. In the next section we will consider the loss of neutrons through the surface of the volume.

You will remember from the previous chapter that when a nucleus undergoes fission, several neutrons can be released. These neutrons would then continue the chain reaction by splitting more nuclei. In the diagram below, you will see an example in which three neutrons (on the right of the diagram) are released by the fission of a single nucleus:

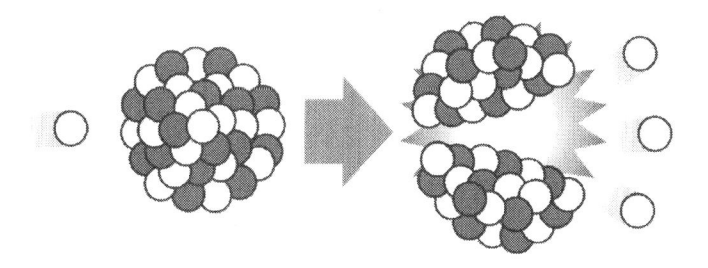

If we are to generate an ever-increasing chain reaction then the question of how many neutrons are released in a single fission is crucial. Clearly, if the chain reaction is to build in intensity then we will need a situation in which the total number of neutrons in the volume increases.

Considering the example of the previous diagram, what is the net increase in the number of neutrons due to the fission of a single nucleus? We can see that one neutron comes in from the left and is absorbed by the nucleus. The nucleus then fissions and three neutrons are released. The net increase in neutrons is therefore going to be three neutrons minus one neutron (the neutron which was initially absorbed), giving a net increase of two neutrons.

On that basis, we can now calculate the build-up of neutrons within the volume of fissionable material. Let us define the average number of neutrons produced per fission as ν (the Greek letter "nu"). In the previous example, ν would be equal to three. Then the net increase in the number of neutrons because of the fission of one nucleus will be given by ν-1, the "-1" term coming from the absorption of the initial incoming neutron. That means that as long as more than one neutron is produced by the fission of a nucleus, then the total number of neutrons will increase over time (ignoring the loss of neutrons through the surface of the volume).

If there is currently a total of N neutrons in a volume of material, and we assume that each of those neutrons goes on to fission a nucleus, then the new total number of neutrons in the volume after the round of fissions will be the net increase in neutrons due to a single fission, multiplied by the number of fissions:

$$(\nu - 1) \times N$$

If the average time between fissions is given by τ (the Greek letter "tau") then the rate of increase in the number of neutrons will be given by:

$$\frac{(\nu - 1)}{\tau} N$$

This formula then tells us the rate of increase of neutrons within a volume of the material – ignoring the loss of neutrons through the surface of that volume. This is the first result we need on our quest to calculate the critical mass.

The derivative

Now let us start to consider the flow of neutrons out through the surface.

The flow of neutrons out through the surface represents a *current* flow of neutrons – just like an electric current is a flow of electrons. And just as the magnitude of an electric current depends on the potential difference (voltage) over a distance, so the current flow of neutrons depends on the difference in the number of neutrons over some distance. In other words, the neutron current depends on the *gradient* of the number of neutrons.

We are well aware of the idea of a gradient as representing the steepness of a hill. In that case, if we are driving up a hill, the gradient is defined as the change in our vertical height divided by the distance we drive in the horizontal direction. This is shown in the following diagram:

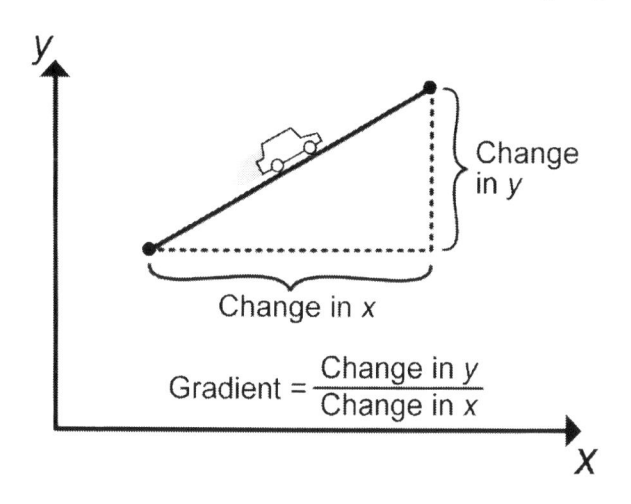

$$\text{Gradient} = \frac{\text{Change in } y}{\text{Change in } x}$$

But the principle of a gradient can be applied to the change of any variable – not just height. We have seen that a gradient is defined as being "the amount of change of a variable divided by the amount of change of another variable". If we have two variables, x and y, then the change of y with respect to x (i.e., the gradient) is known as the *derivative* and is written as:

$$\frac{dy}{dx}$$

You can see how this relates to the gradient at a point:

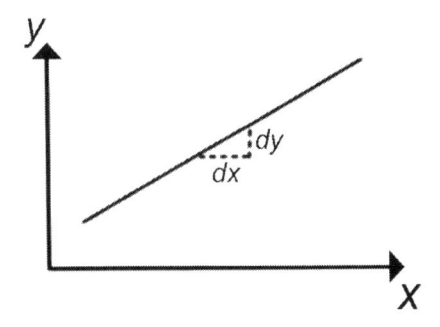

Mathematically, this derivative notation can be used to describe the flow of neutrons out of the material. The value of the critical mass can then be calculated by setting the flow of neutrons out of the material equal to the build-up of neutrons within the material. The lengthy and detailed calculation is contained in the Appendix at the end of this book.

If you want to know the mathematical details of how the world's greatest physicists constructed the world's first atomic bomb, I can highly recommend that you read the Appendix.

The cut a long story short ...

To cut a long story short, the detailed calculation of the value of the critical mass which is described in the Appendix reveals that the critical radius of uranium-235 is 9.1 centimetres. To give an impression of this size of a critical mass, I made my own "critical mass" of the correct dimensions (from a rubber ball and a can of silver spray paint) which you can see me holding in the following photograph:

If this was real, this sphere of metal would explode with a force of several thousand tons of TNT, taking most of my hometown with it.

If you were to hold an actual solid sphere of uranium of this size, the overriding impression would be due to its extraordinary weight. Uranium has the atomic number of 92, which means it has 92 protons in its nucleus – the most of any naturally-occurring element. As most of the mass of an atom is found in its nucleus, this explains why uranium is so heavy. The "critical mass" of uranium I am holding in my hand in the previous photograph would weigh 52 kilograms, or over eight stone. In other words, it would weigh more than many people weigh!

We simply never encounter extraordinary substances such as uranium in our everyday lives. How can this metal weigh so much? How can a small ball of this metal explode with such force? Sometimes these metals glow because of their radioactivity, or they are permanently warm. The properties of these substances seem uncanny and alien. Which raises the question ... how do we obtain these unearthly metals?

6

OAK RIDGE

As explained in Chapter One, it is possible for the nucleus of a particular element to have a different number of neutrons, in which case the substance is called an *isotope*. As isotopes of elements have different numbers of neutrons they therefore have different values for mass number, the mass number being the total number of protons and neutrons in the nucleus. As an example, the isotope of uranium called uranium-238 has a mass number of 238 (which is composed of the sum of 92 protons and 146 neutrons).

Natural uranium, extracted from uranium ore, is composed of two isotopes: uranium-238 and uranium-235 (from now on, uranium-235 will be referred to as "U-235" and uranium-238 will be referred to as "U-238", where "U" is the chemical symbol for uranium). U-238 is by far the most common isotope representing over 99% of natural uranium. However, for purposes of constructing nuclear explosives, it is U-235 which is more interesting.

In this chapter, we will be considering how to obtain U-235 by separating it from U-238 in a process known as *enrichment*.

Remember it was described in Chapter One that neutrons have to act like glue to hold protons together to stop the nucleus from flying apart due to electrical repulsion between the protons. As we have just calculated, U-238 has 146 neutrons, whereas you can similarly calculate that U-235 has the fewer number of 143 neutrons. As the job of neutrons is to act like glue to hold the nucleus together, those three fewer neutrons mean that a nucleus of U-235 is not so strongly "glued together". As a result, U-235 is much more unstable than U-238, and is therefore much more fissionable.

In order to quantify the difference in fission behaviour between U-235 and U-238 we need to introduce the concept of a nuclear *cross section*. Using the usual sense of the term, a cross section is a slice through an object. If we consider a spherical atomic nucleus (or any spherical particle) then a slice through the middle of that particle is going to be a circle. We can then interpret that circle as a target, as shown in the following diagram. The behaviour of the nucleus when it is hit by a bombarding neutron will then depend on whether that neutron hits the target or not:

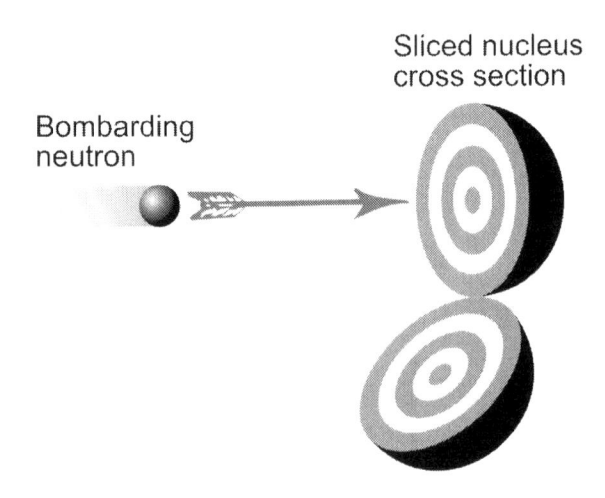

The area of the target is called the nuclear cross section. The different ways in which a nucleus might behave when it is hit by a neutron can be described by defining different cross sections (different sizes of target) for each different type of behaviour. If a cross section is smaller, then that means that the target is smaller, and the behaviour is less likely (the neutron is less likely to hit the target). Conversely, if a cross section is larger, then that means that the corresponding behaviour will be more likely.

As an example, when an atomic nucleus is hit by a neutron, one of three things might happen. Firstly, the neutron might bounce off the nucleus in a process known as *scattering*. The probability of this occurring would be defined by the *scattering cross section* of the nucleus. Secondly, the neutron might be absorbed (captured) by the nucleus. The probability of this occurring would be defined by the *capture cross section* of the nucleus. Thirdly, the neutron might cause the nucleus to split. The probability of this occurring would be defined by the *fission cross section* of the nucleus.

The difference in behaviour when a nucleus of U-235 and a nucleus of U-238 is hit by a neutron can be described in terms of their cross sections. As was explained earlier, a nucleus of U-235 is more fissionable. It can therefore be said that it has a larger **fission** cross section. However, remember that U-235 represents less than one percent of natural uranium, the rest being U-238. This represents a problem if we are trying to achieve a chain reaction in natural uranium because U-238 has a very large **capture** cross section. If we consider a neutron emerging from a fissioned nucleus, we need that neutron to travel unhindered until it reaches a secondary nucleus which it can fission. If, instead, it hits a nucleus of U-238 then the large capture cross section of that nucleus means that the neutron will be captured by that U-238 nucleus without fissioning that nucleus. The chain reaction will then grind to a halt.

Because of this "poisoning" effect of U-238, the only way to achieve an explosive chain reaction is to laboriously separate the U-235 from the much larger proportion of U-238, and then to use only that extracted U-235 in the atomic bomb core.

As stated earlier, less than 1% of natural uranium is U-235, whereas weapons-grade uranium needs to be enriched to 90% U-235. It is therefore clear that a tremendous amount of uranium ore would need to be enriched to make a bomb. The necessary process would resemble an industrial-scale operation more than a laboratory experiment. In 1939, Niels Bohr insisted "it can never be done unless you turn the United States into one huge factory."

Well, effectively, that is just what happened …

The secret city

The scale of the Manhattan Project was staggering. Apart from the development facility at Los Alamos, the project involved more than thirty sites across the United States, Britain, and Canada. These sites included two vast factories in the United States for the production of fissionable material for the bomb cores.

The main industrial facility was the 60,000-acre isotope separation plant (for separating U-235 from U-238) which was located in rural Tennessee. At the time, the area was virtually uninhabited, so the factories and the entire town to house the employees had to be built from scratch. The town was given the rural-sounding name of Oak Ridge in order to make it sound like a small, uninteresting village which would not attract outside interest. In 1945, after being in existence for just two years, Oak Ridge employed 80,000 people and had become the fifth-largest town in the state of Tennessee – and it consumed more electricity than New York City.

At its peak, the Manhattan Project employed 130,000 people and was larger than the entire American automobile industry. It was simply the greatest scientific and engineering gamble in the history of the world. Niels Bohr's prediction that the United States would have to be turned into "one huge factory" was becoming true.

Oak Ridge had to provide all the facilities of a normal town. It had over seven thousand houses, two high schools, nine shopping areas, churches, movie theatres, and a hospital. However, the need for security meant that the entire town had to be surrounded by a barbed-wire fence, and Oak Ridge did not appear on any maps until 1949. To its inhabitants it was called the "Secret City".

To all intents and purposes, Oak Ridge appeared like a normal town, though none of its inhabitants had any idea of why they were doing the jobs they were doing. The difficulty of working in such secrecy was described by one of the managers at Oak Ridge:

> *Well it wasn't that the job was tough – it was confusing. You see, no one knew what was being made in Oak Ridge, not even me, and a lot of the people thought they were wasting their time here. It was up to me to explain to the dissatisfied workers that they were doing a very important job. When they asked me what, I'd have to tell them it was a secret. But I almost went crazy myself trying to figure out what was going on.*

All the workers were sworn to secrecy, and were told never to speak a word to outsiders about their work at Oak Ridge. They were even told what to say if anyone outside asked about their work. If they were asked what they were making in Oak Ridge, they were told to say: "Oh, about 76 cents an hour." And if they were asked how many people

HIDDEN IN PLAIN SIGHT 8

were working in Oak Ridge, they were told to say: "Oh, about half of them."

Here are some photographs of everyday life in the secret city of Oak Ridge during the war:

OAK RIDGE

The largest building in the world

The isotope separation factories at Oak Ridge were faced with a daunting challenge. In his graphic novel version of the Manhattan Project called *Trinity*, Jonathan Fetter-Vorm describes the challenge by using an analogy: "To make a bomb, you need to enrich the uranium, skimming off the less reactive isotope and concentrating the amount of U-235. To get some idea of how hard this is, imagine mixing together two different colours of clay, then trying to separate the colours from each other."

The problem is made more difficult because the relatively simple and cheap process of chemical separation cannot be used. Chemical separation relies on differences in the chemical properties of the two substances being separated: the substances would behave differently in a chemical reaction. Those chemical properties are determined by the distribution of electrons in the shells of the atom, and the numbers of those electrons are, in turn, determined by the number of protons in the atomic nucleus. But two different isotopes of an element have the same number of protons (the same atomic number), so they would have the same distribution of electrons. Therefore, their chemical properties would be the same and chemical separation could not be used.

The only possible alternative is to rely on the difference in the physical properties of the two isotopes, which means the extremely slight difference in the mass of the atoms due to the additional neutrons. On that basis, it is possible to separate the isotopes by using *diffusion*. Diffusion is considered in the Appendix as a means for calculating the critical mass (by considering the diffusion of neutrons through an atomic bomb core), but it is stated in the

Appendix that diffusion describes the spread of any group of particles through a substance (for example, the spread of dust particles in air, or the spread of a chemical through a liquid). In Oak Creek, diffusion was used to separate isotopes by considering the diffusion of uranium atoms through a membrane peppered with microscopic holes.

Firstly, in order to allow diffusion of the uranium, the uranium was turned into a gas: uranium hexafluoride. The gas was then pumped through the membrane. Only gas molecules containing the lighter U-235 isotope would diffuse through the holes in the membrane, so those molecules could be separated from the heavier U-238.

But the very small difference in the atomic mass of the two isotopes meant that it was impossible to achieve perfect separation with a single filter. In fact, the amount of percentage enrichment achieved with each filter was very small. As a result, a high level of enrichment was only possible by using a large number of stages cascaded together, with the output of one stage passing to the input of the next stage. Each stage only provided a very small amount of percentage enrichment, but with enough stages the material became progressively more enriched as it passed through all the stages.

The design of the isotope enrichment process using gaseous diffusion which was used at Oak Ridge was ingenious and is shown in the following diagram:

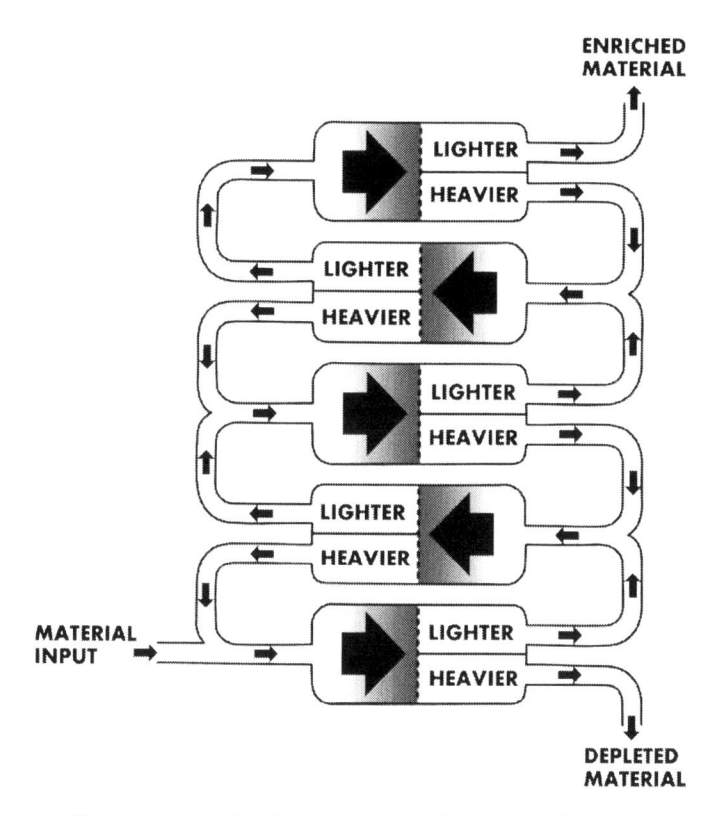

You can see that in each stage the material is pumped against the filter membrane which separates the material into lighter and heavier isotopes. By following the small arrows, you can follow the progressive enrichment of the material on the diagram as it passes through the stages, emerging at the top as highly-enriched material.

You will see that even if some of the material fails to pass through the filter, that material is not wasted. Instead, as you can see on the diagram, the rejected material just drops down to the stage below and is processed again. This is because the filtering process could never be perfect with such a small difference in mass between the isotopes, so the rejected material might still include some atoms of the lighter isotope which should not be wasted.

Because of the very small percentage amount of enrichment achieved by each stage, the extraordinary total of 2,892 stages were required. And these were not small stages: each diffusion tank held 1,000 gallons. As a result, the building which contained the sequence of stages was truly monumental. When it was completed, the Gaseous Diffusion Plant at Oak Ridge was a four-storey U-shaped building which was half a mile long and a quarter of a mile wide. It surpassed the Pentagon to become the largest building in the world – even though the world did not know it existed:

The Calutron Girls

The other major factory at Oak Ridge employed a completely different method of isotope separation based on the electromagnetic force. The technique, known as electromagnetic separation, considered the deflection of electrically-charged uranium atoms as they passed through a magnetic field.

The deflection of the particles is described by *Fleming's left-hand rule for motors*, which you might have been taught in your physics class in school. The rule is described by the following diagram:

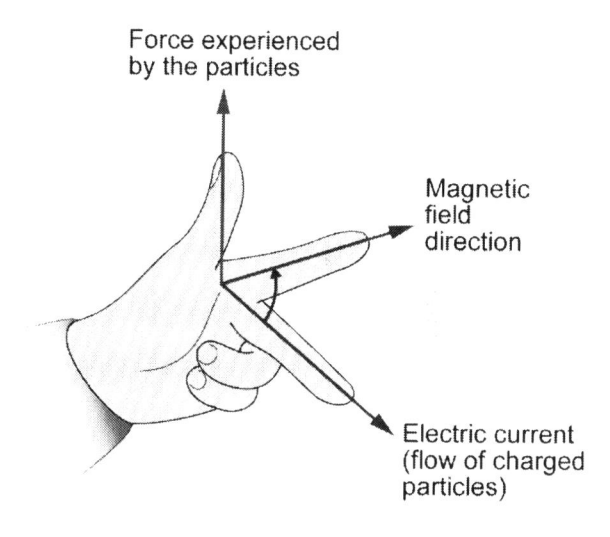

Force experienced by the particles

Magnetic field direction

Electric current (flow of charged particles)

The previous diagram shows that if your middle finger points in the direction in which the charged particles are moving (in the case of a motor, this would be electrons moving down a wire), and if your index finger is pointing in the direction of the magnetic field, then the charged particles will experience a force in the direction of your thumb. This principle was used by the electromagnetic separation process in Oak Ridge to curve the path of beam of uranium atoms as they moved through a magnetic field.

In the following diagram, you will see the charged particles emitted from the source at the bottom of the diagram. At that point, with the direction of the magnetic field into the page, your index finger should point into the page. Your middle finger should always point in the direction in which the particles are travelling (which would be toward the right at the start). In which case, your thumb represents the direction of the force experienced by the particles, which – as you can see on the diagram – will always point toward the centre of curvature:

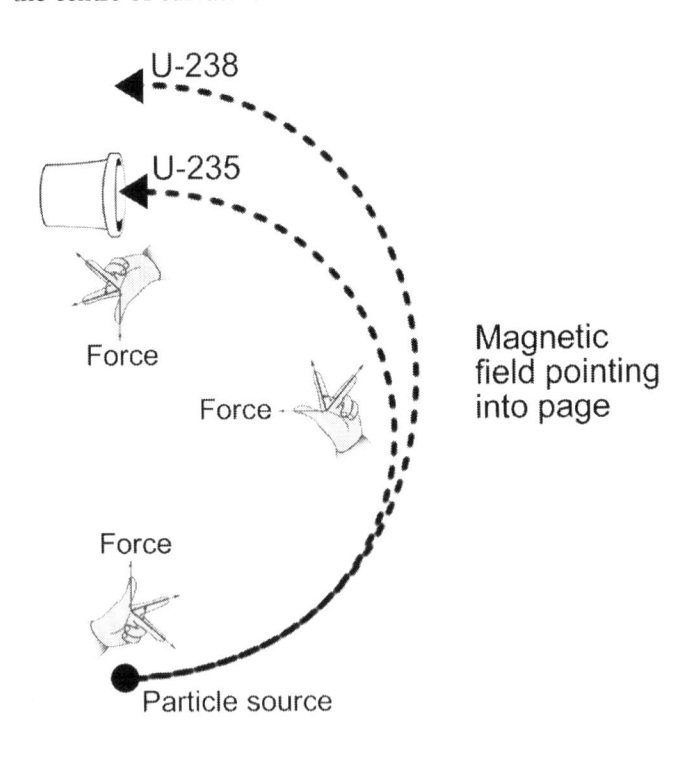

In fact, at every point of their path, the particles will always experience a force toward the centre of curvature, which will result in the beam of particles taking a curved path.

Because the U-235 atoms have slightly lower mass, they will be slightly more affected by the force. In other words, their path will curve more. If a receptacle is correctly placed (a bucket is shown on the diagram) then this will be able to collect only the U-235 atoms and reject the U-238 atoms.

This electromagnetic isotope separation device was called a *calutron*. The name derives from a combination of the words "Cal" and "tron". The "tron" comes from the word *cyclotron* which was an electromagnetic particle accelerator developed by Ernest Lawrence. The "Cal" comes from "California" as Lawrence invented the cyclotron at the University of California.

It can be seen that this calutron method of isotope separation involves building-up the enriched uranium atom-by-atom. It is clear that this would be a very low-yield and time-consuming process: each calutron could produce only 100 milligrams of U-235 per day. In order to increase the yield up to 100 grams of U-235 per day, nine buildings were built at Oak Ridge to house 1,152 calutrons – another monumental operation.

Operating a calutron was a labour-intensive process requiring constant human intervention. In practice, the operators were young women from the local Tennessee community – many straight out of high school or fresh off the farm – who were given the name the "Calutron Girls". Let us now consider the story of the Calutron Girls as described in Denise Kiernan's bestselling book about the female workers of Oak Ridge called *The Girls of Atomic City*.

Many of the young female workers at Oak Ridge had to work in a state of constant anxiety as their brothers and husbands were fighting a brutal war on foreign shores. Despite their anxiety, the women worked tirelessly, motivated by being told that their mysterious work on these futuristic machines would make the war shorter and bring their loved-ones home sooner.

This was the case for Dorothy Jones (affectionately known as "Dot") who was a young girl from Hornbeak, Tennessee. All three of Dot's brothers were fighting in the war, including Shorty who was 23 years old and a deck gunner for the Navy on the battleship USS Arizona. Unfortunately, the Arizona was one of the ships in dock in Pearl Harbor when the Japanese bombed it in December 1942. Dot's family had received the news just before Christmas that Shorty was missing, presumed dead. Dot was determined to do something for the war effort in Shorty's memory, so she had signed up to be one of the Calutron Girls at Oak Ridge.

When Dot entered the building which housed the calutrons, she found it to be a very daunting experience. The building was enormous, with a high ceiling, bright lights, and constant loud electrical sparking sounds from the high-voltage calutrons. Also, bear in mind that in 1944 these were the most futuristic and advanced industrial machines anywhere in the world.

Dot took her place on the wooden stool in front of her calutron control panel, and started her day's work.

The main job assigned to the Calutron Girls was ensuring that the magnetic field strength was set precisely so that the beam of U-235 atoms would be correctly directed into the final receptacle. This was a labour-intensive process as it requiring continuous twisting of handles to adjust the strength of the magnetic field and thereby keep the beam correctly aligned. Here is a photograph of one of the Calutron Girls turning the handles while keeping her eyes firmly fixed on the dial reading:

The following picture shows the Calutron Girls, each woman seated in front of her individual calutron:

Of course, due to the tight security, none of the girls knew the true nature of the task they were performing. The woman seated at the front right of the previous photograph was Gladys Owens who only realised the job she had done during the war when she saw herself in the previous photograph when she toured a museum of Oak Ridge fifty years later.

The girls were warned never to venture behind their control panels because that was where the giant magnet used to generate the powerful calutron magnetic field was positioned. The girls knew that if you carelessly wandered into the magnetic field it would pull the metal hairclips right out of your hair. Or, if you had a belt buckle, it could even pin you against the wall. Legend had it that a maintenance man who had nails in his shoes had been rooted to the spot.

It is clear that the tireless and focussed Calutron Girls performed an essential role in the war effort. However, not everyone was so impressed by the girls. After Ernest Lawrence – the inventor of the calutron – had visited the plant he ran into the office of the chief engineer and blustered: "How dare you pick these silly hillbilly girls to run my machines!" Lawrence insisted that only his male PhD physicists had the ability to run a calutron correctly. Lawrence was certain that the performance of these young girls – straight out of school or off the farm – would be inferior to that of his highly-educated men. However, the chief engineer defended his girls loyally, and told Lawrence how productive the girls had been. He persuaded Lawrence that there should be a production race: Lawrence's highly-qualified physicists versus the Calutron Girls. Whoever produced the most U-235 would be the winner and would get to keep their job. Lawrence reluctantly agreed to the race, totally confident that his men would win.

Of course, in predictable Hollywood movie style, when the race started the Calutron Girls blew the physicists out of the ballpark by collecting a far larger amount of U-235 (apparently the rather geeky physicists were constantly being distracted by getting their slide rules out and analysing any minor variation in the dials). The Calutron Girls were victorious – and they finally won their much-deserved respect. Of course, they were allowed to continue doing their fine job.

The story of the Calutron Girls is one of many remarkable stories of the efforts of the workers at Oak Ridge.

Amazingly, at the end of every week, all the enriched uranium produced by Oak Ridge was taken to Los Alamos **in a briefcase** by a plain-clothed military security officer **on a public train**, with the briefcase handcuffed to his wrist. All actual bomb design and manufacture took place in Los Alamos. It was there that the received uranium was machined to form the bomb cores.

After the dropping of the atomic bomb on Hiroshima in 1945, the nature of the work at Oak Ridge was no longer a secret. The Under Secretary of War, Robert Patterson, sent a letter to all the workers of Oak Ridge:

> *Today the whole world knows the secret which you have helped us keep for many months. I am pleased to be able to add that the warlords of Japan now know its effects better, even than we ourselves. The atomic bomb which you have helped to develop with high devotion to patriotic duty is the most devastating military weapon that any country has ever been able to turn against its enemy. No one of you has worked on the entire project or knows the whole story. Each of you has done his own job and kept his own secret, and so today I speak for a grateful nation when I say congratulations, and thank you all. We are proud of every one of you.*

Mail-order uranium

In this chapter we have examined how the enriched uranium for a nuclear explosive can be produced. It is clear that this is not a simple process. And for that we should be grateful: it is only the difficulty in obtaining suitable nuclear fuel which prevents terrorists and rogue states from making their own nuclear weapons.

However, I thought it would be interesting to discover how difficult it actually is to obtain uranium. Especially when you consider the subtitle of this book is "How to make an atomic bomb". So I did an online search and found it is surprisingly easy to obtain uranium ore.

There are many online companies who will provide you with uranium ore – especially in the U.S. The company **unitednuclear.com** provides a wide range of radioactive material and scientific equipment. Material and equipment is also available on Amazon and eBay. I bought an excellent sample of radioactive uranium ore from eBay.

I also obtained a Geiger counter in order to examine the uranium ore. You can buy many expensive modern small plastic Geiger counters, but I went to eBay and bought an ex-military Geiger counter for a very reasonable price (£50). The Geiger counter was built for the Polish Army in 1970. I get the impression the market is flooded with ex-military equipment, very reliably-built, the Geiger counter will probably last for another hundred years.

The Geiger counter is called a DP-66. There are plenty of these available on eBay. I also bought an external microphone for my DP-66 (to make the clicks audible) from the company **anythingradioactive.com**. They also sell a range of Geiger counters.

A similar American Geiger counter was built around the same time as the DP-66 and it is called the CDV-700. The CDV-700 Geiger counters are also available on eBay and Amazon.

I made a five-minute video of my examination of the uranium ore using my Geiger counter. You can watch the video at the following link:

http://tinyurl.com/radioactivevideo

Here are some stills from the video:

7

PLUTONIUM

In the previous chapter, it was explained that approximately 99% of natural uranium is the isotope U-238 with less than 1% being the desirable fissionable U-235. It was described how that large proportion of U-238 acts to "poison" any chain reaction by capturing stray neutrons. Hence, natural uranium cannot be used in nuclear explosives.

It was then described how the aim of enrichment is to increase the amount of U-235 up to 90%, at which point the enriched uranium is capable of sustaining a chain reaction and is considered to be "weapons-grade" uranium.

On that basis, it might be imagined that U-238 is nothing more than a nuisance, only having the effect of poisoning any chain reaction. However, this would not be a correct assessment of the value of U-238. In fact, U-238 has been used to create more modern nuclear weapons than has U-235.

So how can this be the case? How can U-238 be used to create nuclear weapons? Well, it is because the U-238 is not directly used as an explosive. Instead, it is used to create another element which is highly-fissionable: *plutonium*. It is that plutonium which can be used as a nuclear explosive. As

Richard Rhodes says in *The Making of the Atomic Bomb*, using apparently-useless U-238 to create plutonium "would thereby indirectly put U-238 to work".

To understand how U-238 can be used to create plutonium, remember the discussion of beta decay in Chapter Two. It was explained how beta decay occurs when there are "too many" neutrons in a nucleus, with the result that a neutron converts into a proton – with the emission of an electron, which is the beta particle. The conversion of the neutron into a proton increases the atomic number by one (one more proton in the nucleus) while leaving the mass number unchanged (total number of protons and neutrons is unchanged).

So, in a nutshell, the effect of beta decay is to increase the atomic number by one, thereby creating an element one position further up in the periodic table.

Next, remember in the previous chapter it was explained that U-238 acts to poison a chain reaction because it has a very large capture cross section, which means it is very likely to capture and absorb any neutron which hits its nucleus. If U-238 captures a neutron, it turns into U-239 (one extra neutron increases the mass number by one). The resultant U-239 nucleus would then have an excess neutron, and beta decay would therefore be likely. As just described, the effect of that beta decay would be to move the U-239 one place further up the periodic table, converting it to an element called *neptunium*.

The first line of the following diagram shows a nucleus of U-238 capturing a neutron to become U-239. Follow the curved arrow to see that U-239 then beta decays to neptunium (shown to have an atomic number of 93, which you will see is one more than the atomic number of uranium, which is 92):

$$^{238}_{92}U + ^{1}_{0}n \rightarrow ^{239}_{92}U$$

Neutron

$$^{239}_{92}U \xrightarrow{\text{Beta decay}} ^{239}_{93}Np \xrightarrow{\text{Beta decay}} ^{239}_{94}Pu$$

Neptunium, Plutonium

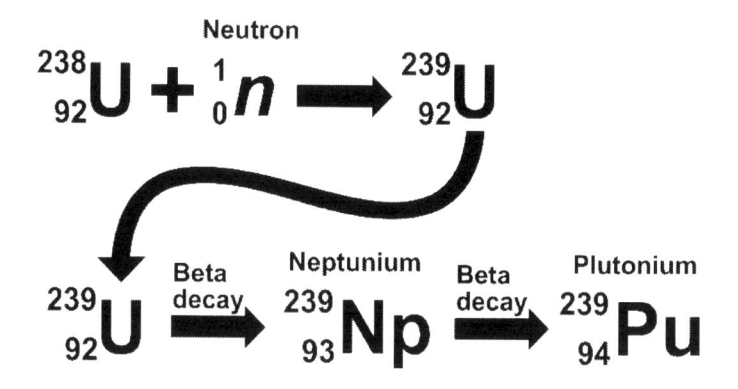

Neptunium is unstable with a half-life of only two days so, as you can see on the diagram, the neptunium quickly beta decays to produce a new element – plutonium – which has an atomic number of 94.

It is interesting to consider the origin of the names "neptunium" and "plutonium". Uranium was originally named after the planet Uranus (which is itself named after the Greek god), which is the seventh planet from the Sun in the Solar System. When the element neptunium was discovered it was decided to continue this naming convention, so neptunium was named after the planet Neptune which is the eighth planet of the Solar System, one place beyond Uranus (Neptune was also named after the Roman god). When plutonium was discovered, the naming convention was continued so it was named after Pluto which, at the time, was considered to be the ninth planet of the Solar System just beyond Neptune (rather ominously, Pluto was also the name of the Greek god of death).

The following diagram shows the three outermost planets of the Solar System, together with the three elements named after them. The consecutive atomic numbers of those elements (92, 93, 94) are also shown:

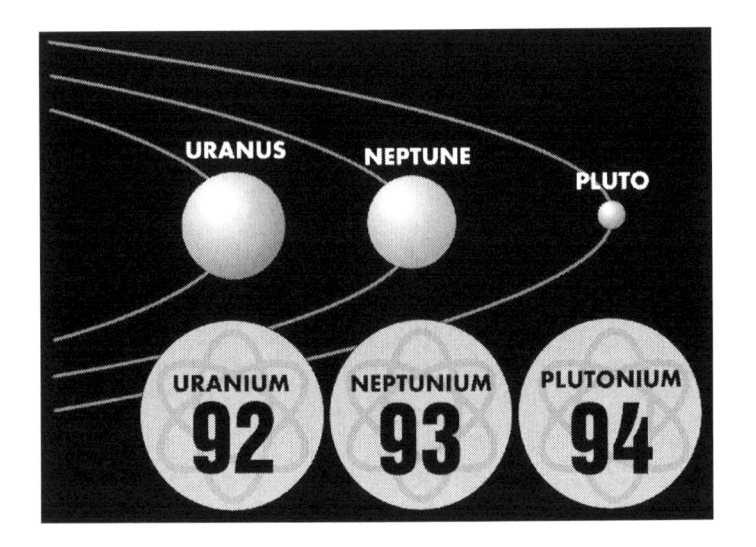

Uranium and plutonium are the only elements which can be used as nuclear explosives. So why is it the case that uranium and plutonium are nuclear explosives, but neptunium is not? After all, as can be seen, neptunium lies between uranium and plutonium in the periodic table. To understand why only uranium and plutonium are nuclear explosives, we need to consider the concept of *parity* in the atomic nucleus.

In mathematics, parity indicates whether a number is even or odd. If we consider the number of protons and neutrons in an atomic nucleus, it is known that the parity of those numbers plays a large role in describing the fission behaviour of that nucleus.

Firstly, if we consider uranium-235, it has an atomic number of 92. That means it has 92 protons, which is an **even** number. It therefore must have 143 neutrons (235-92) which is an **odd** number. So the parity of U-235 is **even/odd**.

If we then consider neptunium in a similar manner, we find it has 93 protons and 146 neutrons, so the parity of neptunium is **odd/even**, which is clearly different to the parity of fissionable U-235.

However, if we apply the same reasoning to plutonium, we find it has 94 protons and 145 neutrons, which is **even/odd** parity, **which is the same as fissionable U-235.** So this is the reason why plutonium fissions in a similar manner to U-235, and why plutonium can also be used as a nuclear explosive.

The nuclear reactor

If it takes the capture of a single neutron by a single nucleus of U-238 to produce plutonium then it is clear that to produce a significant amount of plutonium would require trillions of neutrons. In practice, the only way to generate those neutrons would be by starting a self-sustaining chain reaction in uranium, with each fissioned uranium nucleus producing more than one neutron.

But it has already been explained how U-238 acts to poison any chain reaction because of its tendency to capture any free neutrons. So how can U-238 be involved in any chain reaction in uranium?

The answer is that we have to slow the neutrons down.

It might seem counter-intuitive that slowing-down neutrons can encourage a chain reaction based on the fissioning of nuclei – it might be suspected that faster neutrons would be more effective at splitting nuclei. However, a slower neutron can be much more effective. This discovery was made by Enrico Fermi, and Fermi's discovery was described by Richard Rhodes:

> *Everyone had assumed that faster neutrons were better for nuclear bombardment because faster protons and alpha particles always had been better. But the analogy ignored the neutron's distinctive neutrality. A charged particle needed energy to push through the nucleus's electrical barrier. A neutron did not. Slowing down a neutron gave it more time in the vicinity of the nucleus, and that gave it more time to be captured.[6]*

This effect was particularly notable in the isotope U-235 rather than U-238. For U-235, the fission cross section for slow neutrons is absolutely huge. It other words, if you have a sample of U-235 and you bombard it with slow neutrons, the nuclei will almost certainly fission, rather than the nuclei capturing those neutrons. In fact, it is that huge fission cross section of U-235 for slow neutrons which allows a chain reaction to be generated in natural uranium (i.e., uranium which has not been enriched). This is despite U-235 making up less than 1% of natural uranium. The remarkable fission cross section of U-235 allows a chain reaction in natural uranium **as long as the neutrons are slowed down**. And this is what happens in a *nuclear reactor*: a chain reaction is generated in natural uranium by slowing down the neutrons.

In order to slow the neutrons down in a nuclear reactor, the uranium has to be surrounded by a material called a

---

[6] This effect can also be understood in a different way. Because we are dealing with particles, quantum mechanical effects dominate. Every particle has an associated radius called the *de Broglie wavelength* which can be understood as being the cross section of a point particle. A slow-moving neutron will have low energy and will see a larger de Broglie wavelength of the nucleus, which means a larger fission cross section.

moderator. The neutrons collide with nuclei in the moderator and lose energy: they slow down. An example of a moderator which is used in nuclear reactors is graphite, which is a form of carbon. A fast neutron which has been emitted from a fissioned nucleus needs to travel through 20 centimetres of graphite, colliding with graphite nuclei on its way, before its speed is sufficiently reduced. So it is clear that a nuclear reactor has to be quite a large structure. In the reactor core, spheres or rods of natural uranium are interspersed with the moderator material, which acts to slow down the neutrons.

So a nuclear reactor is based on a nuclear chain reaction in much the same way as we have seen a nuclear bomb is also based on a nuclear chain reaction. But, because slow neutrons are involved rather than fast neutrons, the energy released by a nuclear reactor is slow and controllable rather than fast and uncontrolled (as in a nuclear bomb). This is also the reason why slow neutrons (and natural uranium) cannot be used to manufacture a nuclear bomb: a nuclear bomb requires fast neutrons and enriched uranium.

And if a nuclear reactor is based on a self-sustaining nuclear reaction, then it has to have a critical mass of material – again, just like in a nuclear bomb. But the inclusion of large amounts of moderator material, and the complex arrangement of fuel rods, made it difficult to calculate the value of that critical mass – unlike the relatively simple critical mass calculation for a nuclear bomb core which was presented in Chapter Five of this book. Hence, the value of the critical mass for a nuclear reactor had to be found from experiment. Richard Rhodes considers the problem:

> *A slow-neutron chain reaction in natural uranium, like its fast-neutron counterpart U-235, requires a critical mass: a volume of uranium and moderator sufficient to sustain neutron multiplication despite the*

inevitable loss of neutrons from its outer surface. No one knew the specifications of that critical volume, but it was obviously vast – on the order of some hundreds of tons.

It was clear that in order to calculate the value of the critical mass, and to prove that a nuclear chain reaction was possible, a test reactor would have to be built.

The world's first nuclear reactor

We are so used to the idea of nuclear reactors producing electrical power that it may come as a surprise that the very first nuclear reactor was designed without the slightest interest in its potential to produce power. It has just been explained that a chain reaction in natural uranium in a nuclear reactor is the perfect environment to produce plutonium for nuclear bombs, and it was for this purpose that the first nuclear reactor was built.

To tell the story of the world's first nuclear reactor, we must once again catch up with the itinerant Leo Szilard. When we last left Leo Szilard in 1939 he was just a guest researcher in Columbia University in New York City for three months. Amazingly, in November 1940, Szilard had managed to find himself a proper job at last, and had been added to the Columbia payroll. Szilard was worked hard with Enrico Fermi throughout 1941, trying to create an experimental nuclear reactor. They had been trying to generate a slow-neutron chain reaction in natural uranium, using graphite as a moderator. So far, they had not been successful.

With the attack on Pearl Harbor in December 1941, and the launch of the Manhattan Project, a decision was taken in January 1942 to centralise all reactor research at the

University of Chicago. Hence, Leo Szilard was on the move once again.

It was clear that the reactor would be extremely large and would require a lot of space. The University of Chicago was very supportive of the war effort, and agreed that the physicists could use any of the extensive sports facilities. There were a number of rackets courts under the West Stand of the university's football stadium at Stagg Field. A decision was made to locate the reactor in a doubles squash court.

In his successive reactor experiments, Fermi had to determine how close he was to achieving a sustained chain reaction. In order to measure his progress toward that goal, Fermi calculated a value for the average number of neutrons produced per fission in the reactor. As explained in Chapter Five of this book (the calculation of critical mass), as long as more than one neutron is produced by the fission of a nucleus, then the total number of neutrons will increase over time and a self-sustaining chain reaction will result. The goal for Fermi, therefore, was to ensure that the average number of neutrons produced per fission was more than 1.0. However, in his first attempt at creating a reactor, Fermi calculated that neutron production number to be a disappointing 0.87.

By May 1942, Fermi had improved the purity of the uranium and graphite in his experimental reactor and had managed to improve the neutron production value to 0.995. While this was still short of the value of 1.0 required for a chain reaction, it was believed that further improvements to the purity of the materials would now push the value beyond 1.0. Therefore, at that point, work began on the full-scale reactor.

A total of 45,000 black graphite blocks were required. They were all machined to a standard size approximately the size of a shoebox. The reactor was built in layers on the floor of the squash court. The first layer was purely made of graphite blocks, containing no uranium. The second layer was made of blocks which had two seven-centimetre diameter uranium spheres embedded. The layers alternated in this fashion as the reactor was built up.

The resultant arrangement of the graphite blocks is shown in the following diagram. You can see that some of the blocks have two spheres of uranium embedded in them, whereas some of the blocks have no uranium spheres. The end result is that the uranium spheres form a perfect cubic lattice, with each sphere surrounded by the necessary 20 centimetres of graphite moderator:

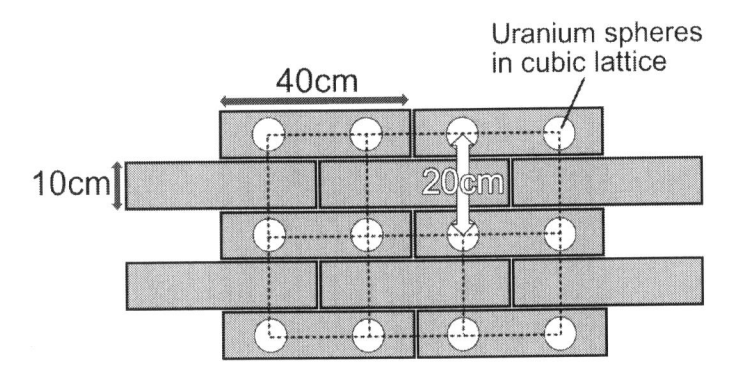

Ten control rods made of cadmium were used to keep the reaction under control. The rods were thirteen foot long and were inserted into the reactor. Cadmium has a huge capture cross section for neutrons, so the control rods could be used to dampen the chain reaction if necessary and keep it within safe bounds.

For security reasons, no photographs were taken of the Chicago reactor, but we do have an artist's rendering. At the top of the ladder, you can see one of the control rods poking out of the blocks of graphite:

After 57 layers had been completed, Fermi's measurements of neutron intensity indicated that the reactor would now go critical if all of the control rods were removed. By this point, the reactor was 20 feet high and 25 foot wide. The reactor was completed on December 1st 1942, but it was decided that the experiment would take place the following day.

The temperature the following morning was below zero. A crowd of about forty people gathered on the balcony of the squash court to witness the historic event. As each of the control rods were removed slowly, the neutron intensity increased but then levelled off – the reactor was still subcritical. The last remaining control rod was then withdrawn slowly, with Fermi predicting "Now it will become self-sustaining. The trace on the recorder will climb and continue to climb; it will not level off." Fermi's colleague, Herbert Anderson, describes what happened next:

At first you could hear the sound of the neutron counter, clickety-clack, clickety-clack. Then the clicks came more rapidly, and after a while they began to merge into a roar. Suddenly Fermi raised his hand and said that the pile had gone critical. No one present had any doubt about it.

Now the chain reaction had started, the neutron intensity would keep rising instead of levelling-off. If the control rods had not been reinserted then after an hour-and-a-half the rate of increase would have resulted in a power output of a million kilowatts. But before it would have reached that stage, it would have killed everyone in the room and melted through the concrete in the floor and the underlying rock. Theoretically, there is nothing that can stop a reactor in meltdown before it melts its way to the Earth's core.

The Chicago event was historic. It was the first artificial nuclear chain reaction, the birth of nuclear power, and the first time appreciable energy had been released from the atom. Crucially, it also opened the way to the production of plutonium as a nuclear explosive.

As described earlier in Chapter Four, just eight years earlier, Leo Szilard had crossed the road in Southampton Row in London and had first had the idea of how a nuclear chain reaction might be achieved. Standing on the balcony of a squash court in Chicago, he had now seen his idea become a reality.

The Demon Core

Plutonium is an incredibly dangerous substance. Plutonium is extremely toxic if it is ingested or inhaled, and the radiation from plutonium can also be deadly. The American environmental activist Ralph Nader once claimed that a pound of plutonium could kill eight billion people.

A single plutonium core (sphere of plutonium) was responsible for two separate horrific fatal incidents at Los Alamos. The sphere of plutonium had been created to be used in a series of "criticality experiments" designed to accurately determine the necessary critical mass of plutonium for the bomb. The sphere was only 3.5 inches in diameter – about the size of an orange. It was designed to be 5% smaller than the critical mass, and therefore safe for experimentation. However, that was a small safety margin, and if any neutrons which were emitted by the core were reflected back into the core then that could cause the core to become critical.

On August 21st 1945, the physicist Harry Daghlian was working alone, performing a criticality experiment using the core. When Daghlian accidentally dropped a tungsten carbide block near the core, the neutrons which were reflected by the block turned the core critical. There was a sudden blue flash – and even the air seemed to glow blue (this was because of the fluorescence of the air molecules when they are hit by charged radioactivity). Daghlian quickly threw the brick away, but it was too late: he died 25 days later from radiation poisoning.

Precisely nine months later, Louis Slotin was working with the same core, performing a different criticality experiment. However, Slotin's dangerous technique was unapproved. Slotin had surrounded the core with two hemispheres of neutron-reflecting beryllium. Slotin was keeping the hemispheres apart by using the blade of a screwdriver as a lever. Slotin was full of bravado, and had performed the experiment a dozen times before to various teams of observers, proud of his ability to wield his screwdriver. When Enrico Fermi saw Slotin's experiment he told Slotin that he would be "dead within the year".

The following photograph shows a re-creation of Slotin's experiment showing the hemispheres being kept apart by a screwdriver blade (the spherical core is hidden inside the hemispheres, approximately the same size as the two spheres you can see on the right):

On the day of the accident, Slotin's screwdriver slipped and the two hemispheres came together for a fraction of a second. There was the usual flash of blue light showing that the core had gone critical. The criticality lasted for only half a second before the core expanded due to the heat produced, and was then no longer critical. But the damage was done.

Slotin died nine days later.

As a result of these two fatal incidents by the same core, the core was given the ominous name the "Demon Core". The second of the Demon Core incidents was dramatised in the 1989 movie *Fat Man and Little Boy*. The character of Louis Slotin in the movie was played by John Cusack. The Demon Core sequence is accurate and extremely chilling. The sequence is available on YouTube at the following link:

http://tinyurl.com/thedemoncore

Right at the start of the sequence if you look carefully you can see the two hemispheres being kept apart by nothing more than the blade of Slotin's screwdriver, and you can even see the small spherical plutonium core inside the hemispheres. You can later see Slotin twisting the wooden handle of his screwdriver to move the upper hemisphere up and down. Then, as Slotin's screwdriver slips, you can see the deadly flash of blue light.

After these two fatal incidents, all hands-on criticality experiments at Los Alamos stopped. From then on, remote-controlled machines were used to perform the experiments, with all personnel safely watching TV cameras a quarter of a mile away.

8

DETONATION

In this final chapter we will move on to consider the final step of the process: actual bomb design, and how to detonate a nuclear bomb.

The sex of the bomb

The message from the discussion so far seems to be that the theory behind building a nuclear bomb is fairly straightforward: you just have to assemble a critical mass of fissionable material in order to generate an explosive chain reaction. However, there is another major technical challenge involved in the detonation of a nuclear bomb: there is a need to assemble the critical mass in a very short period of time.

The critical mass must be assembled from smaller sub-critical pieces. A problem arises in that before the critical mass is fully assembled, there is inevitably a short period of time when the total assembled mass is smaller than the critical mass. During this very short window of time, it is possible for a small chain reaction to start in the material. This small chain reaction might be initiated by an external source, for example, a neutron from a cosmic ray. Only a very small amount of the material will be fissioned and the bomb will detonate with a vastly smaller explosive power. This small explosion will blow the rest of the bomb apart before the majority of the material is fissioned. This represents a failure of the nuclear explosive. This problem is called *predetonation*, and the resultant relatively small explosion is called a *fizzle*.

In order to avoid predetonation, it is essential that the entire critical mass is assembled in an extremely short period of time. In practice, this needs to be less than one millisecond.

The simplest way in which a critical mass can be rapidly assembled is to split the critical mass into two sub-critical pieces, and then fire those two pieces together at great speed by using a conventional explosive. An example is presented in the following diagram which shows an assembly mechanism which has been successfully used in a nuclear bomb. The diagram shows a hollow tube of U-235 being fired down a gun barrel towards a bullet of U-235. When the two pieces of U-235 are joined together, the resultant mass will be equal to (or greater than) the critical mass and the bomb will explode:

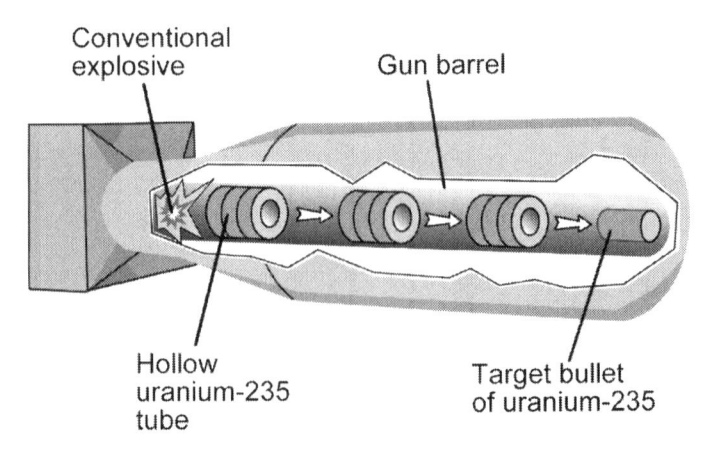

It might appear rather perverse to fire the hollow tube of uranium onto the bullet, rather than performing the reverse operation by firing a "male" bullet into the hollow tube. In fact for fifty years after the first atomic bomb was constructed it was always believed that the bullet was fired into the tube, and this was always how it was presented in every history book. However, we now know the correct design of the bomb thanks to the investigative work of a truck driver from Wisconsin named John Coster-Mullen.

Coster-Mullen, together with his son Jason, has built a perfect replica of the Hiroshima atomic bomb. The construction details were laboriously obtained from a huge variety of different sources including interviews with machinists who worked on the bomb, and by forensic examination of photographs of the bomb.

In 1994, as part of his detective work to uncover the secrets of the bomb, Coster-Mullen had a telephone interview with one of the bomb's original engineers named Harlow Russ. Russ was being careful not to divulge classified information, but in the middle of the interview he happened to mention: "You know the projectile was hollow, didn't you?" In other words, Harlow Russ had revealed that a hollow tube was fired onto the bullet – not vice versa as had been assumed. Coster-Mullen had discovered that the "sex" of the bomb was female – not male.

According to the journalist Howard Morland: "Every encyclopaedia in the world, from the Britannica to the World Book, described how the Hiroshima bomb was made, and included a diagram. News articles and school teachers referenced these diagrams. But here's the thing: every single one got it wrong. John Coster-Mullen and his self-published memoir got it right."

The story of John Coster-Mullen – and a discussion about the sex of the bomb – is available at the following link:

http://tinyurl.com/costermullen

Coster-Mullen has also published a book called *Atom Bombs* which contains design details of the first atomic bombs. The details are so accurate that this is truly "the book that the Feds tried to ban". In 2005, Coster-Mullen was handed an internal Los Alamos email that described discussions taking place at the highest levels of the Los Alamos National Laboratory to identify a way to quash his book. The book is also permanently banned from sale at the National Museum of Nuclear Science and History in Albuquerque on orders from Sandia National Laboratory. According to Harold Agnew, the former Scientific Director of Los Alamos: "There are drawings in there that are absolutely correct. He's got everything exactly: dimensions, materials, and things that have been really classified. His cross-section drawings are the most incendiary portions of his book."

Here is a photograph of me looking suitably awestruck at the cross-sectional diagrams in my copy of John Coster-Mullen's book *Atom Bombs*:

Implosion

Let us now calculate how long it takes to assemble the critical mass using this method. The following calculation is from Bruce Cameron Reed's book *The History and Science of the Manhattan Project*.

The gun muzzle velocity will be approximately 1,000 metres/second, and the lengths of both the projectile tube and the target cylinder will be approximately 10 cm. These measurements are shown on the following diagram:

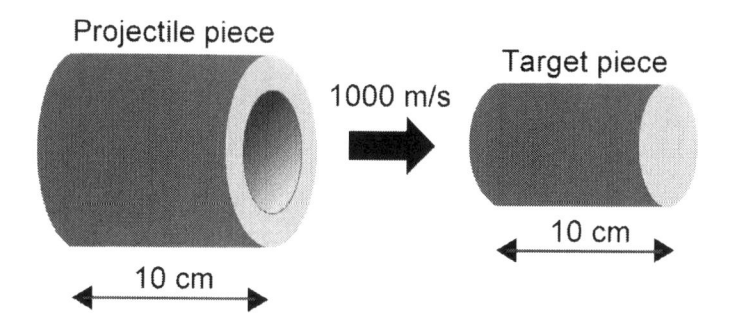

Projectile piece **Target piece**
1000 m/s
10 cm
10 cm

At that speed of travel, it will take the projectile piece about 100 microseconds to completely surround the target piece. So, using the "gun assembly" mechanism, it is possible to assemble a critical mass in approximately 100 microseconds. Hence, this method can be used to create a critical mass of uranium (it was stated earlier that assembly time for a critical mass of uranium needs to be less than one millisecond).

However, it was discovered that the problem of predetonation was more serious for a plutonium bomb. The problem arises because plutonium was found to have a far higher rate of spontaneous fission than uranium. Spontaneous fission can occur when a nucleus splits randomly of its own accord, without being hit by a neutron.

Because of random neutrons being produced by spontaneous fission, it was likely that the chain reaction would start before the projectile piece had completely surrounded the target piece, which would result in a much smaller "fizzle" explosion. This meant that the previously-described gun assembly mechanism taking approximately 100 microseconds would be too slow to detonate a plutonium bomb. As an alternative, a more sophisticated "implosion" method was developed at Los Alamos.

In order to understand how the implosion method works, let us remember the principle which underlies the calculation of the critical mass. As described in Chapter Five, a critical mass of material is created when the amount of neutrons escaping through the surface area of the material is less than the amount of neutrons being generated within the material. So the obvious way to create a critical mass is to add more material. But it is clear that an alternate method would be **to reduce the surface area of the material**, and thereby reduce the number of neutrons escaping. This can be achieved by compressing the material, and increasing its density in the process.

The implosion method works by surrounding the bomb core with conventional high-explosive. The resultant explosion can have a huge compressive effect, pushing inward at a speed of several kilometres a second and achieving pressures comparable to those at the Earth's core. It takes only about a microsecond to compress the core to criticality, which is clearly far faster than the 100 microseconds it takes for the gun assembly method. So the implosion method avoids the predetonation problem of plutonium.

However, it is extremely difficult to compress a heavy metal object in a symmetrical manner using implosion. When the method was first proposed, there were many in Los Alamos who thought it was not possible. The problem was compared to the problem of crushing a full beer can without

DETONATION

spilling any beer. A brilliant young physicist called Richard Feynman who was working at Los Alamos considered the implosion method and simply said "It stinks".

The main problem arises because simply smearing a uniform spherical coating of high-explosive around the bomb core will not produce a smoothly symmetrical compression of the core. To understand why that is the case, consider the following diagram:

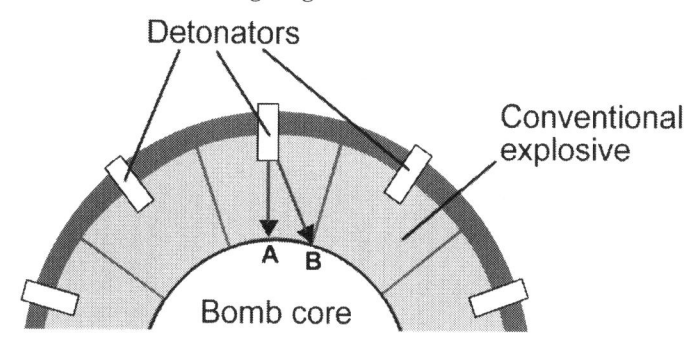

The diagram shows a uniform coating of conventional explosive (shaded in light grey) spread evenly around a bomb core. You can see that a series of rectangular detonators are inserted into the explosive at a regular spacing.

In the diagram, let us just consider the top detonator. Because of the curvature of the bomb core, you can see on the diagram that the distance from that detonator to a point A on the surface of the bomb core is slightly shorter than the distance from the detonator to a point B on the surface of the bomb core. What this means is that the wave of detonation of exploding material spreading from the detonator will reach point A slightly earlier than it will reach point B. As a result, the compression will be applied earlier to point A and the result would be to deform the bomb core – like crushing and deforming a beer can in your hand – rather than applying a smooth compression at all points at the same time.

Clearly, the key to achieving a smooth compression is to ensure that the detonation wave hits all points on the surface of the bomb core at the same time. This can be achieved by using a combination of fast-burning explosive and slow-burning explosive, as shown on the following diagram:

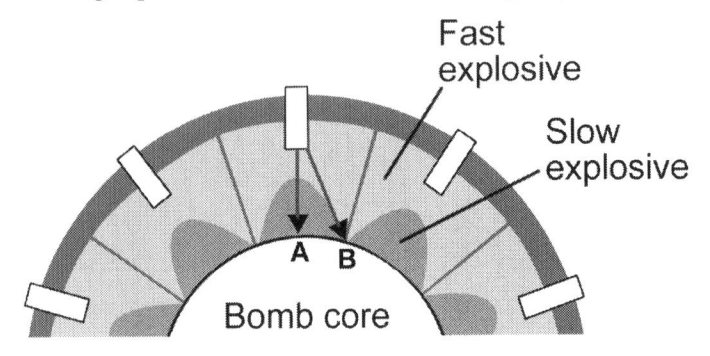

Fast explosive

Slow explosive

Bomb core

The previous diagram shows two different types of conventional explosive being used, which burn at different speeds. The fast-burning explosive is shown in the lighter grey than the slow-burning explosive.

Because of this arrangement of fast-burning and slow-burning explosives, you will see that even though the distance from the top detonator to point A is shorter, the detonation wave will have to pass through more of the slower explosive. Whereas the distance from the top detonator to point B is longer, but the detonation wave will only pass through the fast explosive to get to point B. As a result, the detonation wave reaches both points A and B at the same time, applying equal compression to all points at the same time. The result is a smooth compression of the bomb core.

These arrangements of explosive charges have the effect of "bending" the emerging shockwave from the detonator so that the shockwave meets the entire surface of the core at the same time. In other words, the shockwave is bent in order to fit the shape it is intended to squeeze. These

DETONATION

arrangements of explosives are sometimes called explosive lenses because of the similarity to the way an optical lens bends a light wave, based on the similar principle that light travels at different speeds through different media.

Interestingly, in modern implosion bombs the surrounding explosives are often modified by random amounts so that the timing for individual detonators has to be correspondingly modified. The timing device is then stored away from the bomb. In this way, even if a bomb is stolen by a terrorist group they will not be able to explode it without the corresponding custom timing device. This technology – called a Permissive Action Link (PAL) – was developed by the U.S. in the 1960s and freely-given to the Soviet Union to make the world a safer place.

So that basically describes the structure of a plutonium bomb based on implosion. However, there are two structures that have not been described yet, and these are shown on the following diagram:

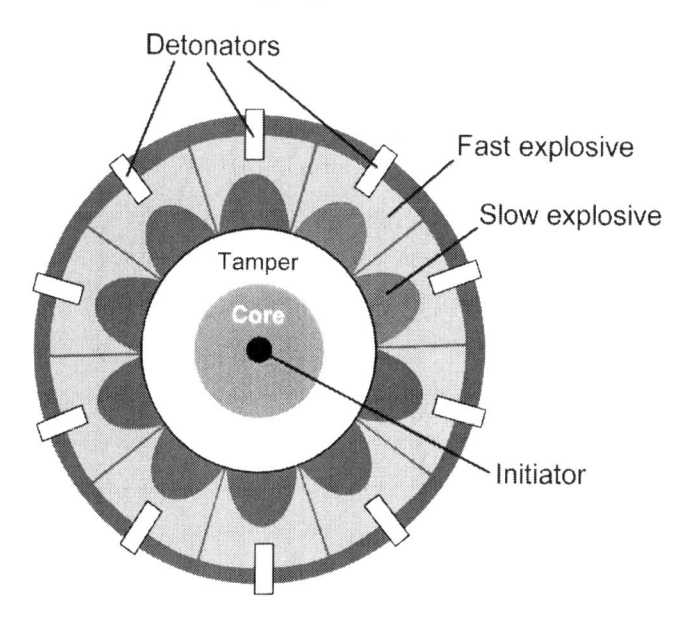

Around the plutonium core can be seen the *tamper*, which is a jacket of heavy metal. This acts to increase the yield (the explosive power) of the explosion, and it does this in two ways. Firstly, the sheer weight of the tamper restricts the expansion of the core due to the extreme heat generated in the core, thus giving the core more time to fission completely. It is important that the core does not expand too rapidly as once its surface area grows too large it no longer represents a critical mass and the chain reaction stops. The tamper also acts to reflect neutrons back into the core and stop them escaping through the surface of the core.

At the centre of the core you will see a small black sphere, about the size of a golf ball. This is called the *initiator*, and it is a neutron source which is designed to produce the very first neutron of the chain reaction.

The initiator is usually made of a small piece of the radioactive element polonium surrounded by some beryllium, which is a very light metal. This arrangement is known to provide a good source of neutrons. Energetic alpha particles from the polonium can knock neutrons out of the beryllium nuclei. When the core is crushed, the beryllium mixes with the polonium sending out a shower of millions of neutrons – enough to kick-start the chain reaction.

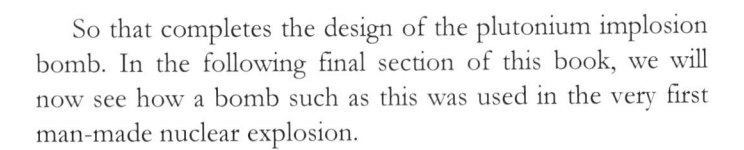

So that completes the design of the plutonium implosion bomb. In the following final section of this book, we will now see how a bomb such as this was used in the very first man-made nuclear explosion.

Trinity

In the middle of the barren New Mexico desert, 160 miles south of Los Alamos, there stands a black stone obelisk, about 12 foot high. The obelisk is made of local lava-rock. There is a plaque on the obelisk on which just a few words are written: "Trinity Site Where the World's First Nuclear Device Was Exploded on July 16, 1945."

The obelisk marks the hypocentre of the explosion.

There are several reasons why the site was chosen. The site had to be flat to allow for accurate measurements of the explosion, the site had to have little wind to prevent the spread of radioactive fallout, and it was preferred to have a site reasonably close to Los Alamos.

The following map shows the location of the Trinity test site. You can see it lies about 160 miles south of Los Alamos, and about sixty miles away from the town of Alamogordo:

Los Alamos — Trinity

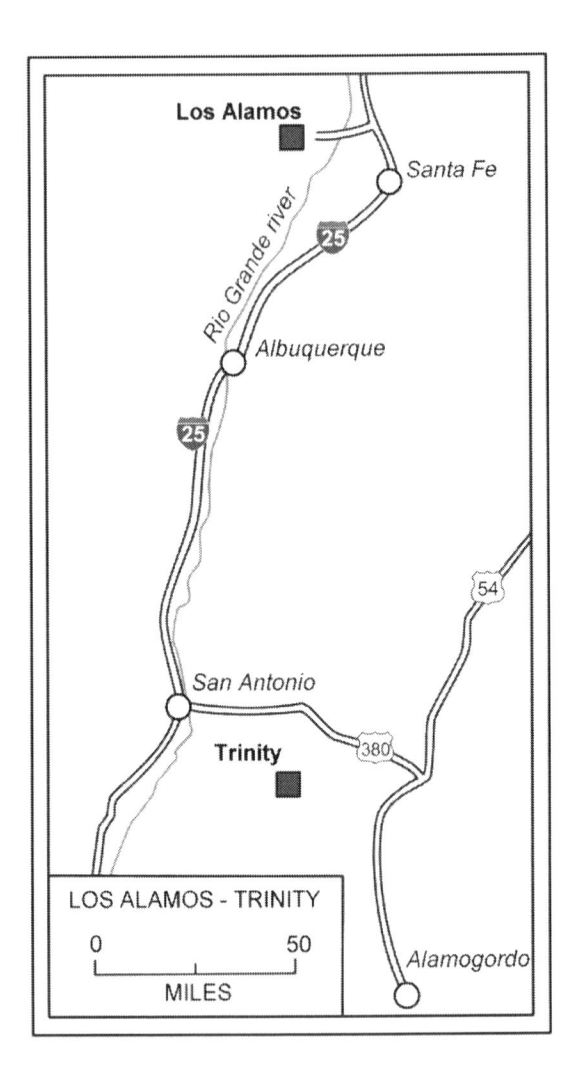

0 — 50 MILES

The decision was taken that a test of the plutonium bomb was essential, because of the uncertainties about the implosion mechanism. The plutonium had cost about 250 million dollars to produce, and if the test failed then the plutonium could be recovered and re-used.

DETONATION

The plutonium core and the surrounding shell of high explosive were manufactured at Los Alamos and transferred separately to the Trinity test site. The plutonium core left Los Alamos first on the afternoon of Thursday July 12th (the test was planned for the early morning of the following Monday). The orange-sized piece of metal was given the VIP treatment, driven to Trinity on the back seat of an army sedan with a convoy of armed guards riding in front and behind. The high explosive was driven overnight to avoid traffic and potential accidents, taking eight hours to travel at thirty miles an hour to Trinity.

At Trinity, a temporary base camp was constructed nine miles away from where the bomb was to be detonated. The following photograph shows the Trinity base camp:

The bomb was to be detonated at the top of a steel tower, one hundred feet tall. There was an oak platform at the top of the tower which could support those working on the bomb, and there was a hole in the platform through which the bomb could be raised. The platform was surrounded on three sides by corrugated iron, making a shack. Here is a photograph of the Trinity tower:

At 1pm on the Friday afternoon, the high explosive shell arrived at the tower. The plutonium core arrived later in the afternoon. Final assembly of the bomb took place in a canvas tent at the base of the tower. The bomb assembly group was headed by Norris Bradbury. It was Bradbury who finished the bomb assembly in the late evening under floodlights.

The following morning, the Saturday, the canvas was removed and the bomb was hoisted to the top of the tower by an electric winch. When the bomb was at the top of the tower, the detonators were inserted into the bomb. Here is a photograph of Norris Bradbury standing by the fully-assembled bomb at the top of the tower:

DETONATION

The power cables you can see attached to the bomb are attached to the detonators which were spaced around the high-explosive shell. The power cables were so thick because a new type of detonator was developed specifically for the plutonium bomb. If you remember back to the discussion about implosion earlier in this chapter you will remember the importance of the shockwave hitting all points of the spherical core at the same time in order to achieve smooth compression of the core. However, all that work on explosive lenses would have been pointless if all the detonators did not all detonate at precisely the same time.

In order to achieve that microsecond accuracy of detonation, a new type of detonator was developed called an exploding bridge wire detonator. The detonator was made of a thin wire through which a pulse of extremely high current was passed. The current needed to be in the region of 100 kiloamps, which explains the large power cables attached to the bomb. The wire instantly becomes hot and literally explodes – starting the detonation. Crucially, all the detonators will detonate within a microsecond when the pulse of high current is applied.

The detonators were inserted on the Saturday, so Sunday became a day of waiting before the planned detonation of the bomb early on Monday morning. An intense thunderstorm hit the site at 2 a.m. on Monday morning. According to Isidor Rabi: "It was raining cats and dogs, lightning and thunder. We were really scared that this object there in the tower might be set off accidentally. So you can imagine the strain on Oppenheimer." The detonation was delayed until just before dawn. A telephone call was made to the governor of New Mexico warning him that he might have to declare martial law if the explosion caused a panic.

At 5:30 a.m., the weather had cleared. The detonation sequence was controlled from inside an earth-covered concrete control centre six miles from the bomb tower. Oppenheimer would watch the explosion from that shelter.

Busloads of visitors from Los Alamos were ready to view the explosion from the summit of Compania Hill, twenty miles from the bomb. The crowd included many of the physicists who had worked on the bomb. It was dark and cold, and the tension was almost unbearable. They had been advised to lie face down on the ground, with their feet pointing toward the explosion. They had also been given welding glasses to protect their eyes. However, Richard Feynman refused the glasses, later claiming to have been the only person to have viewed the explosion with no eye protection. According to one observer: "It was an eerie site to see a number of our highest-ranking scientists seriously rubbing sunburn lotion on their faces and hands in the pitch-blackness of the night, twenty miles from the expected flash."

In the control centre just six miles from the bomb, the atmosphere was so tense that Oppenheimer could hardly breathe. He held onto a post to steady himself and stared directly ahead as the last seconds ticked away.

5 ... 4 ... 3 ... 2 ... 1 ... 0 ...

DETONATION

First came the instant blast of light. Pitch black night suddenly turned to the most intense daylight, three times brighter than the brightest Sun. Those lying on the ground at Compania Hill, twenty miles from the explosion, saw distant hills suddenly brightly illuminated as if by a rising Sun.

Isidor Rabi at Base Camp, nine miles from the explosion, described the light:

> *We were lying there, very tense, in the early dawn. Those ten seconds were the longest ten seconds that I ever experienced. Suddenly, there was an enormous flash of light, the brightest light I have ever seen or that I think anyone has ever seen. It seemed to last forever. You would wish it would stop; altogether it lasted about two seconds. Finally it was over, and we looked toward the place where the bomb had been; there was an enormous ball of fire which grew and grew and rolled as it grew; it went up into the air, in yellow flashes and into scarlet and green. It looked menacing. It seemed to come toward one.*
>
> *A new thing had just been born; a new control; a new understanding of man, which man had acquired over nature.*

Philip Morrison described the other overwhelming sensation of the radiant heat:

> *The thing that got me was not the flash but the blinding heat of a bright day on your face in the cold desert morning. It was like opening a hot oven with the Sun coming out like a sunrise.*

DETONATION

Norris Bradbury explained the difficulty in making sense of something which no one had ever seen before:

Most experiences in life can be comprehended by prior experiences, but the atom bomb did not fit into any preconceptions possessed by anybody.

The bomb had exploded with force of twenty thousand tons of TNT. The temperature at the centre of the explosion was four times greater than the temperature at the centre of the Sun. The resultant pressure was more than a hundred billion times the pressure at the surface of the Earth, and was the greatest pressure ever to exist on the Earth. In Arizona, 150 miles away, it was reported that a woman was puzzled why she saw "the Sun come up and go down again".

The greatest physics experiment of all time had been a success.

Robert Oppenheimer, the director of the Manhattan Project, was in the control centre just six miles from the bomb. His brother Frank, also a physicist, was by his side. After the incredibly tense last few minutes, Oppenheimer could breathe freely at last.

They both walked outside and climbed on top of the earth-covered bunker. Frank turned to face his brother – who within the next few days would become the most famous man in the world – and said "It worked". Oppenheimer looked at the immense mushroom cloud now climbing high over the desert.

"Yes", said Oppenheimer, "It worked".

FURTHER READING

The History and Science of the Manhattan Project
by Bruce Cameron Reed
A highly-detailed and technical description of the science of the atomic bomb.

The Making of the Atomic Bomb by Richard Rhodes
The definitive account of nuclear physics leading up to the development and use of the first atomic bomb.

Trinity by Jonathan Fetter-Vorm
A graphic novel of the development of the atomic bomb, with plenty of good science. Extremely well done.

APPENDIX: CALCULATING THE CRITICAL MASS

This Appendix is highly-mathematical, so only read it if you are confident about your mathematical abilities.

This Appendix contains the detailed calculation of the value of the critical mass, and continues from the discussion in Chapter Five. We will start by calculating the flow of neutrons out of the material. This initial section is going to introduce some mathematical terminology which is likely to be new to you. Please do not worry about it or get bogged-down by the details. All we will need is the result at the end of this section which is a well-established result in mathematics.

The spread of particles through a substance (for example, the spread of dust particles in air, or the spread of a chemical through a liquid) is called *diffusion*. The theory behind diffusion has been well-studied, and there is a branch of physics specifically called *diffusion theory*. We can apply precisely the same theory to consider the flow of neutrons out of a volume (after all, neutrons are just another type of small particle – like dust particles).

To cut a long story short, in this section we will be seeing that diffusion theory tells us that that the flow of neutrons out of the material is described mathematically by the *Laplacian* operator. All we will be needing from this section is to find the formula for the Laplacian operator.

OK, so let's go …

As described in Chapter Five, the value of the neutron current is proportional to the gradient of the number of neutrons. So, using the derivative notation, we can express the value of the neutron current, j, as:

$$j = -D\frac{dN}{dr}$$

where N is the number of neutrons in a volume of the fissionable material, and r represents the distance from the centre of the material. Assuming the fissionable material to be in the shape of a solid sphere, then r represents the distance from the centre of the sphere towards the edge of the sphere.

You will see that there is another constant in the expression: D is a constant called the *diffusion coefficient* which is a fixed property of the fissionable material. We will be calculating the value of D later in this chapter.

The minus sign in the formula arises because the current is from the larger neutron concentration to the smaller neutron concentration, whereas a positive gradient would normally be from low to high (as in climbing a hill, which would represent a positive gradient).

But if the value of the neutron current remains the same over a distance then the number of neutrons entering a volume of the material (the current in) will be the same as the number of neutrons leaving the volume of the material (the current out). In other words, the number of neutrons in that volume will be unaltered. But what we are really interested in is the **change** in the number of neutrons in that volume. That will then give us the rate at which the neutrons are spreading into space, escaping out of the volume. Mathematically, this is called the *divergence*.

So if we want to calculate the divergence (the change in the number of neutrons in a volume) then we have to consider the **change** in the value of the neutron current. In other words, we have to consider the derivative of the neutron current. But we calculated earlier that the value of the neutron current, j, is given by:

$$j = -D\frac{dN}{dr}$$

which is, clearly, a derivative itself. So to find the change in the number of neutrons we need to take the derivative of a derivative. Therefore, the change in the number of neutrons in a volume is calculated by the derivative of j:

$$-D\frac{d}{dr}\left(\frac{dN}{dr}\right)$$

What we have done here is show that the change of the number of neutrons in a volume is given by the divergence of the gradient. In mathematics, it is well-known that the divergence of the gradient is called the *Laplacian*. As I said earlier, this is all we need to know from this section. The mathematical form of the Laplacian is well-known and you can even find it on the Wikipedia page for the Laplacian.

The Laplacian for spherical coordinates (which is what we want) is given halfway down the Wikipedia page:

http://tinyurl.com/laplaciansspherical

It does indeed have the two derivatives (a derivative of a derivative), which is what we just calculated. You will see that the precise form is:

$$-D\frac{1}{r^2}\frac{d}{dr}\left(r^2\frac{dN}{dr}\right)$$

Creating the equation

So now we have a formula describing the flow of neutrons out of a volume. We are making good progress.

As stated at the start of Chapter Five, the critical mass occurs when the loss of neutrons through the surface of a volume is equal to the build-up of neutrons inside that volume. In Chapter Five we calculated that the rate of increase in the number of neutrons in a volume of the material is given by:

$$\frac{(\nu-1)}{\tau} N$$

Putting this value equal to the previous equation for the loss of neutrons through the surface of that volume gives:

$$-D \frac{1}{r^2} \frac{d}{dr}\left(r^2 \frac{dN}{dr} \right) = \frac{(\nu-1)}{\tau} N$$

Dividing both sides by D and N, and changing the sign of both sides, gives:

$$\frac{1}{N}\left[\frac{1}{r^2} \frac{d}{dr}\left(r^2 \frac{dN}{dr} \right) \right] = -\frac{(\nu-1)}{D\tau}$$

This equation describes the distribution of neutrons in a critical mass. So now all we need to do is solve the equation and find the shape of the distribution. However, I must admit, the equation does look very complicated! But do not worry – we will find the equation has a surprisingly simple solution.

And in order to solve the equation, we are going to use a quite amazing tool …

The genius of Mr. Wolfram

We now have an equation we need to solve to find the value of N, which will tell us the distribution of the number of neutrons within the fissionable material. You will see that the equation includes the value of N, and also the derivative of N. Equations which include combinations of variables and their derivatives like this are called *differential equations*.

Differential equations can capture the dynamics of a complex system in which elements are moving at different rates to one another. As a result, differential equations are very useful tools for describing physical phenomena such as heat flow or the motion of waves, and are used throughout physics and engineering. But differential equations can also be used to describe any complex system with "moving parts" such as biology and economics.

The **solution** of differential equations then takes a central role in many science problems. The problem can be complex because differential equations can come in a huge variety of different forms. For many simple forms, the method of solution is well known and described in mathematical textbooks. But when more complicated situations are described (for example, modelling the movement of airflow in weather forecasting), the differential equations might have no simple solution and it is then necessary to use a powerful computer to perform repeated calculations, slowly converging to a solution.

Our attempt to calculate the critical mass of an atomic bomb has resulted in the creation of a differential equation, and we now have to solve that equation. However, the task of solving differential solutions just got a whole lot easier thanks to the artificial intelligence website Wolfram Alpha.

The driving force behind Wolfram Alpha is the controversial mathematical genius Stephen Wolfram. Wolfram was a child prodigy who published physics papers at the age of 15. In 1988 he left academia to develop the software package called Mathematica. Mathematica can be used to solve complex problems, and to visualise solutions. The program was an instant success and made Wolfram his fortune.

In 2002, Wolfram released a controversial book called *A New Kind of Science* in which he claimed that great complexity arises from simple rules called *cellular automata*. Wolfram claimed to be able to use these simple programs to model almost any complex system. Unfortunately, the book was widely-criticised for its arrogant tone and unscientific approach, and there was the suggestion that the project represented a huge ego-trip for Wolfram. It would appear that even geniuses have their failings.

In 2008, Wolfram released the Wolfram Alpha search engine (or "computational knowledge engine" as they call it). The search engine technology behind Wolfram Alpha claims to be unique in that it can answer questions which are posed in everyday language. I thought I would test this claim of Wolfram Alpha, so I went to its webpage:

http://www.wolframalpha.com

In the Wolfram Alpha search box I entered the question "Who was the oldest Beatle?" Sure enough, in just a few seconds it came back with the correct answer "John Lennon". In contrast, entering the same query into Google just returns the usual long list of search results. It is clear that Wolfram Alpha is attempting to do something very different: it is trying to **understand** your question, and **calculate** an

APPENDIX

answer (this is opposed to the usual search engine method of just matching character strings). This is true artificial intelligence: a brain on the web.

What makes Wolfram Alpha particularly interesting for our particular problem is that some of the mathematical processing ability of Mathematica has been incorporated into Wolfram Alpha. This makes Wolfram Alpha incredibly useful for solving any mathematical problems you might encounter. For example, I was surfing the net recently when I saw a mathematical puzzle asking me to find the value of x in the following equation:

$$\sqrt{x+15} + \sqrt{x} = 15$$

Can you solve the puzzle and calculate the value of x?

Well, it looked a bit of a brain-ache to me, so I went straight to the Wolfram Alpha website:

http://www.wolframalpha.com

and in the search box at the top of the page, I typed the puzzle in English just as it is written:

square root of (x+15) plus square root of x equals 15

So the search box looked like:

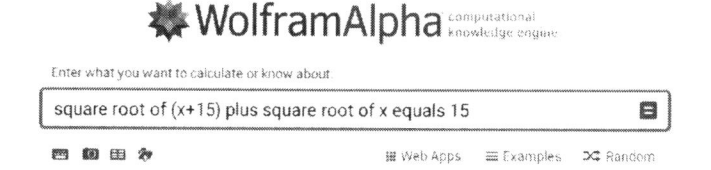

Try it yourself. Then click the icon at the extreme right of the search box to start Wolfram Alpha computing the solution. After a few seconds computation, it returns the correct answer, giving you the value of x (you may have to scroll down the page to see the solution). I won't tell you what the answer is – try it yourself!

So Wolfram Alpha can understand most mathematics problems (or any question) which you type in natural English. That makes it very handy for solving a wide range of puzzles. In fact, in typically modest Wolfram style, it claims to have the ambitious long-term goal of "computing whatever can be computed about anything".

Of particular interest, Wolfram Alpha is able to recognise a differential equation and find its solution – and that's good news for us because we are going to use Wolfram Alpha to solve our differential equation …

APPENDIX

Solving the equation

If you remember, this is our differential equation:

$$\frac{1}{N}\left[\frac{1}{r^2}\frac{d}{dr}\left(r^2\frac{dN}{dr}\right)\right] = -\frac{(\nu-1)}{D\tau}$$

Yes, it looks very complicated, but as I said earlier we will find it has a very simple solution.

I am going to make a temporary modification to this equation purely for reasons of brevity and convenience, and also to make it easier to enter into Wolfram Alpha. You will see that the right-hand side of the equation is formed from a combination of various constants, which are based on the properties of the fissionable material. The end-result will just be another different constant. So, purely so that I don't have to write out the same long expression each time, let us replace the right-hand side of the equation with the constant value, k. In other words:

$$k = \frac{(\nu-1)}{D\tau}$$

In which case, our differential equation then becomes:

$$\frac{1}{N}\left[\frac{1}{r^2}\frac{d}{dr}\left(r^2\frac{dN}{dr}\right)\right] = -k$$

When we finish our calculation, we will replace k with the original combination of constants.

In order to solve this equation, we have to discover the form of the distribution of neutrons which is described in the equation by N, with the value of N varying as we move from the centre of the material to the edge of the material. We are going to use Wolfram Alpha to solve this equation in rather a backward fashion. We are going to suggest a likely candidate solution for N, and then use Wolfram Alpha to check if that potential solution satisfies the differential equation. If that is shown to be the case, then we have a verified solution for our equation and we would have achieved success. However, if the solution is only slightly incorrect then we will need to slightly modify our candidate solution and run it through Wolfram Alpha again.

So what type of candidate solution might we try to describe the distribution of neutrons? Well, we would imagine that the number of neutrons in the material would be highest in the centre of the material as that is the region which is furthest from the edge. Conversely, we would imagine that the number of neutrons would be lowest at the edge of the material as that is where the losses occur. And in between those two extremes we would imagine the number of neutrons would decrease smoothly along a curve.

This suggests a smooth curve like a sine wave, and there is a simple potential candidate solution which fits the bill:

$$N(r) = \frac{\sin(r)}{r}$$

Let us now use Wolfram Alpha as a visualization tool to help us see what this candidate solution looks like. Will it resemble the expected distribution of neutrons in a critical mass? To find out, go to the Wolfram Alpha website:

http://www.wolframalpha.com

APPENDIX

In the search box at the top of the page, enter the following string into the search box (take care to note the first "minus pi" term):

plot (sin(r)/r) from r=-pi to r=pi

Your Wolfram Alpha search box should then look like this:

WolframAlpha computational knowledge engine.

Enter what you want to calculate or know about:

| plot (sin(r)/r) from r=-pi to r=pi |

≡ Web Apps   ≡ Examples   ⤭ Random

Click the icon at the extreme right of the search box to start Wolfram Alpha computing the solution. After a few seconds computation, you will be rewarded with the following graph:

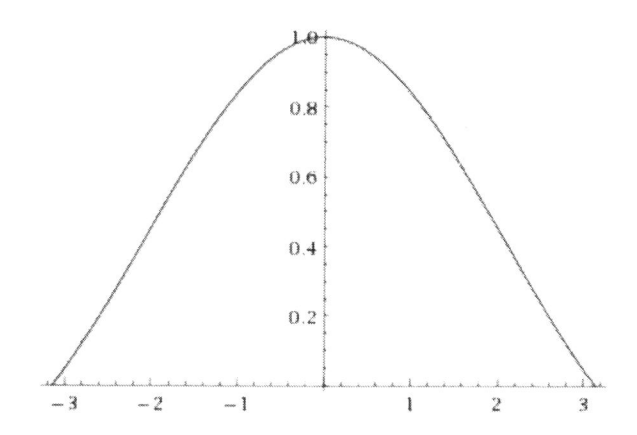

You can see the distribution of neutrons is at its highest in the middle of the graph, which represents the centre of the fissionable material ($r=0$). There is then a smooth decline in the number of neutrons toward the edge of the material as losses are higher nearer the edge. The number of neutrons is zero at the edge as any neutrons produced by fission are free to fly straight out of the material, so there is never any opportunity there for the number of neutrons to build up.

So the general shape of this graph looks much as we would expect. This gives us confidence in our candidate solution.

If you are stuck and having problems getting Wolfram Alpha to work correctly, I have already entered the correct data and generated the correct graph. So, only if you are stuck, the following link should always work, taking you directly to the correct page showing the graph:

http://tinyurl.com/neutrongraph

For our next step, we will use a different feature of Wolfram Alpha: its capability to solve differential equations. To recap, our candidate solution is:

$$N(r) = \frac{\sin(r)}{r}$$

Can this really the solution we are seeking? We might have our doubts as it does seem to be rather simple. We need to check to see if this potential solution satisfies our differential equation. To do that, we need to replace both occurrences of N in our differential equation with $\sin(r)/r$. That gives:

$$\frac{1}{\frac{\sin(r)}{r}} \left[\frac{1}{r^2} \frac{d}{dr} \left(r^2 \frac{d \frac{\sin(r)}{r}}{dr} \right) \right] = -k$$

APPENDIX

We will check if this is the correct solution by using Wolfram Alpha to calculate the value of the left-hand side of the equation, and checking that the result is equal to minus k (the right-hand side of the equation).

So return to the Wolfram Alpha webpage and carefully enter the following expression as a single continuous line into the search box:

1/(sin(r)/r) * 1/(r squared) * derivative of ((r squared) * derivative of (sin(r)/r))

This expression represents the left-hand side of the previous differential equation. Compare the expression with the left-hand side of the differential equation and see if you can spot that the various elements in the two forms correspond to each other.

Your Wolfram Alpha search box should now look like this:

Click the icon at the extreme right of the search box to start Wolfram Alpha computing the solution. After a few seconds computation, it should come back with its answer:

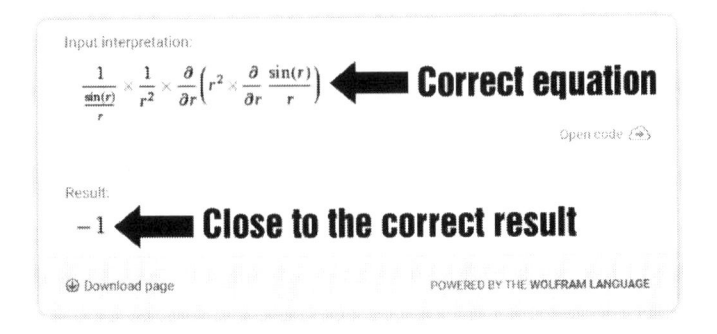

$$\frac{1}{\frac{\sin(r)}{r}} \times \frac{1}{r^2} \times \frac{\partial}{\partial r}\left(r^2 \times \frac{\partial}{\partial r} \frac{\sin(r)}{r}\right)$$ ⬅ **Correct equation**

Result: -1 ⬅ **Close to the correct result**

If everything has gone correctly then at the top of the answer you should see that Wolfram Alpha has correctly interpreted the input string into the correct form of the differential equation. And underneath that you will see that Wolfram Alpha has calculated the value of the expression and found that it is equal to minus one.

If you are stuck and having problems getting Wolfram Alpha to work correctly, I have already entered the correct data and the computation worked correctly for me. So, only if you are stuck, the following link should always work, taking you directly to my results page:

http://tinyurl.com/candidatesolution

The result of minus one is an encouraging result, though it is not quite correct. We were hoping that the result would be equal to the right-hand side of our differential equation, which is minus k. But minus one is very similar to minus k, so all we have to do is tweak our candidate solution slightly.

APPENDIX

Clearly, in order to get a result of minus k, our candidate solution will have to include the value of k somewhere in it. So let us modify our candidate solution slightly by including the value of k in some way. With that in mind, let us try this new candidate solution for the distribution of neutrons:

$$N(r) = \frac{\sin(r\sqrt{k})}{r}$$

You will see that the square root of k has been added to the candidate solution. Let us now check if that satisfies our differential equation (which has now become even more complicated!):

$$\frac{1}{\frac{\sin(r\sqrt{k})}{r}} \left[\frac{1}{r^2} \frac{d}{dr} \left(r^2 \frac{d\frac{\sin(r\sqrt{k})}{r}}{dr} \right) \right] = -k$$

Let us enter the left-hand side of this equation into Wolfram Alpha and see if it gives us the answer we want. So return to the Wolfram Alpha webpage and carefully enter the following expression as a single continuous line into the search box:

```
1/(sin(r * square root of k)/r) * 1/(r squared) *
    derivative of ((r squared) *
    derivative of (sin(r * square root of k)/r))
```

Once again, click the icon at the extreme right of the search box to start Wolfram Alpha computing the solution. After a few seconds computation, it should come back with its answer:

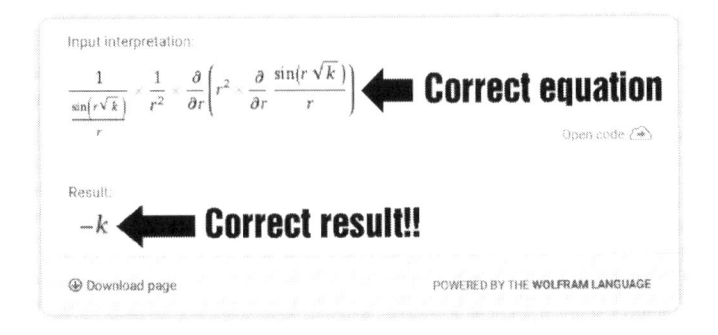

```
Input interpretation:
  1        1    ∂  ⎛     ∂   sin(r√k) ⎞
─────── × ─── ─── ⎜ r² ─── ─────────  ⎟     ⬅ Correct equation
sin(r√k)  r²   ∂r ⎝     ∂r     r      ⎠
───────
   r
                                              Open code

Result:
 −k   ⬅ Correct result!!

 Download page            POWERED BY THE WOLFRAM LANGUAGE
```

Hooray! Wolfram Alpha has calculated that the result is equal to minus k, which is the correct result we were looking for (minus k was the value on the right-hand side of our differential equation). So our complicated differential equation turned out to have a really simple answer: the accurate formula for the distribution of neutrons in our critical mass is given by:

$$\frac{\sin(r\sqrt{k})}{r}$$

If you are stuck and having problems getting Wolfram Alpha to work correctly, I have already entered the correct data and the computation worked correctly for me. So, only if you are stuck, the following link should always work, taking you directly to my correct results page:

http://tinyurl.com/resultspage

If you followed all the steps and everything worked correctly, then congratulations! Now, for the last step, let's put some actual numbers into this result and calculate the value for the critical mass in terms of kilograms.

APPENDIX

Calculating the critical mass

We have now done the hard part of this calculation. The rest is now relatively simple.

To recap, we have shown that the distribution of neutrons in our critical mass is described by:

$$\frac{\sin(r\sqrt{k})}{r}$$

To obtain the value of r when we have a critical mass of the material, we will use our knowledge of the properties of the edge of the critical mass. In solving a differential equation, this is called a *boundary condition*.

We know that the net amount of neutrons produced at the edge will be the smallest value as that is where neutron losses out of the material are largest. From our high-school knowledge of trigonometry we know that the sine function reaches a minimum at π radians (or, equivalently, 180 degrees). Indeed, if you look at our earlier graph showing neutron distribution, you will see that it reaches zero when the horizontal scale is at π radians. So, in the previous formula, with r set equal to the critical radius, R_c, that means the contents of the bracketed term must be equal to π.

So, just considering the contents of the bracketed term:

$$R_C\sqrt{k} = \pi$$

Let us now replace k in this equation with the combination of constants which k represents (remember, we did the reverse substitution earlier). The equation now becomes:

$$R_C \sqrt{\frac{v-1}{D\tau}} = \pi$$

Squaring both sides of the equation gets rid of the square root:

$$R_C^2 \left(\frac{v-1}{D\tau} \right) = \pi^2$$

Rearranging the elements of this equation gives:

$$R_C^2 = \frac{\pi^2 D\tau}{v-1}$$

Which is the formula for the critical radius which you will find in Section 10 of the *Los Alamos Primer* which deals with the calculation of the critical mass.

Let us take the square root of both sides to get our final result for the critical radius:

$$R_C = \sqrt{\frac{\pi^2 D\tau}{v-1}}$$

And that's it! That is the final equation for the critical radius of the critical mass! If you managed to follow all the steps – and completed the Wolfram Alpha exercises on your computer – then you can genuinely tell your friends that you used a computational knowledge engine to solve a differential equation to calculate the distribution of neutrons in a nuclear bomb core. If they are not impressed by that – then get new friends.

APPENDIX

Let us now put numbers into this formula to get a value for the critical mass:

- If you remember it was explained in Chapter Five that ν is the average number of neutrons produced from the fission of one nucleus, which for uranium-235 is equal to 2.52.

- Also explained in Chapter Five, τ is the average time between fissions. This is a very short period of time, only a few nanoseconds. As a result, a nuclear explosion releases all its energy in just a microscopic fraction of a second. To be precise, τ is equal to only 8.1×10^{-9} seconds.

- D, which was introduced earlier is the diffusion coefficient. D is calculated from the average distance between collisions (the *mean free path*) which is 0.029 metres, multiplied by the velocity of the neutrons which is 1.7×10^7 metres/second, divided by the number of dimensions of space which is 3. That means D is equal to 1.6×10^5 metres^2/second.

If you substitute these three values into the previous formula for the critical radius you will get the result that the critical radius of uranium-235 is 9.1 centimetres.

PICTURE CREDITS

All photographs are public domain unless otherwise stated.

Photograph of Ernest Rutherford's laboratory is licensed under the Creative Commons license and is provided by Wikimedia Commons (tinyurl.com/rutherfordlaboratory).

Photograph of the Kaiser Wilhelm Institute is by Peter Kuley and is provided by Wikimedia Commons.

Diagram of moose size in Sweden is by Nmccarthy16 and is provided by Wikimedia Commons.

HIDDEN PLAIN SIGHT 9

The mystery of consciousness

COMING SOON

Printed in Great Britain
by Amazon